Circuit Design with VHDL

Circuit Design with VHDL

Volnei A. Pedroni

MIT Press
Cambridge, Massachusetts
London, England

Contents

Preface

Structure of the Book

The book is divided into two parts: Circuit Design and System Design. The first part deals with everything that goes directly inside the *main code*, while the second deals with units that might be located in a *library* (for code sharing, reuse, and partitioning).

In summary, in Part I we study the entire background and coding techniques of VHDL, which includes the following:

- Code structure: libraries, entity, architecture (chapter 2)
- Data types (chapter 3)
- Operators and attributes (chapter 4)
- Concurrent statements and concurrent code (chapter 5)
- Sequential statements and sequential code (chapter 6)
- Objects: signals, variables, constants (chapter 7)
- Design of finite state machines (chapter 8)
- And, finally, additional circuit designs are presented (chapter 9).

Then, in Part II we simply add new building blocks, which are intended mainly for library allocation, to the material already presented. The structure of Part II is the following:

- Packages and components (chapter 10)
- Functions and procedures (chapter 11)
- Finally, additional system designs are presented (chapter 12).

Distinguishing Features

The main distinguishing features of the book are the following:

- It teaches in detail all indispensable features of VHDL synthesis in a concise format.
- The sequence is well established. For example, a clear distinction is made between what is at the *circuit* level (Part I) versus what is at the *system* level (Part II). The *foundations* of VHDL are studied in chapters 1 to 4, *fundamental coding* in chapters 5 to 9, and finally *system coding* in chapters 10 to 12.
- Each chapter is organized in such a way to collect together related information as closely as possible. For instance, *concurrent code* is treated collectively in one chap-

ter, while *sequential code* is treated in another; *data types* are discussed in one chapter, while *operators and attributes* are in another; what is at the *circuit level* is seen in one part of the book, while what is at the *system level* is in another.

• While books on VHDL give limited emphasis to digital design concepts, and books on digital design discuss VHDL only briefly, the present work completely integrates them. It is indeed a *design-oriented* approach.

• To achieve the above-mentioned integration between VHDL and digital design, the following steps are taken:

  • a large number of *complete* design examples (rather than sketchy or partial solutions) are presented;

  • illustrative top-level *circuit diagrams* are always shown;

  • fundamental *design concepts* are reviewed;

  • the solutions are *explained* and commented;

  • the circuits are always *physically* implemented (using programmable logic devices);

  • *simulation* results are always included, along with analysis and comments;

  • finally, appendices on programmable devices and synthesis tools are also included.

Audience

The book is intended as a text for any of the following EE/CS courses:

• VHDL
• Automated Digital Design
• Programmable Logic Devices
• Digital Design (basic or advanced)

It is also a supporting text for in-house courses in any of the areas listed above, particularly for vendor-provided courses on VHDL and/or programmable logic devices.

Acknowledgments

To the anonymous reviewers for their invaluable comments and suggestions. Special thanks also to Ricardo P. Jasinski and Bruno U. Pedroni for their reviews and comments.

I CIRCUIT DESIGN

1 Introduction

1.1 About VHDL

VHDL is a *hardware description language*. It describes the behavior of an electronic circuit or system, from which the physical circuit or system can then be attained (implemented).

VHDL stands for VHSIC Hardware Description Language. VHSIC is itself an abbreviation for Very High Speed Integrated Circuits, an initiative funded by the United States Department of Defense in the 1980s that led to the creation of VHDL. Its first version was VHDL 87, later upgraded to the so-called VHDL 93. VHDL was the original and first hardware description language to be standardized by the Institute of Electrical and Electronics Engineers, through the IEEE 1076 standard. An additional standard, the IEEE 1164, was later added to introduce a multi-valued logic system.

VHDL is intended for circuit *synthesis* as well as circuit *simulation*. However, though VHDL is fully simulatable, not all constructs are synthesizable. We will give emphasis to those that are.

A fundamental motivation to use VHDL (or its competitor, Verilog) is that VHDL is a standard, technology/vendor independent language, and is therefore portable and reusable. The two main immediate applications of VHDL are in the field of Programmable Logic Devices (including CPLDs—Complex Programmable Logic Devices and FPGAs—Field Programmable Gate Arrays) and in the field of ASICs (Application Specific Integrated Circuits). Once the VHDL code has been written, it can be used either to implement the circuit in a programmable device (from Altera, Xilinx, Atmel, etc.) or can be submitted to a foundry for fabrication of an ASIC chip. Currently, many complex commercial chips (microcontrollers, for example) are designed using such an approach.

A final note regarding VHDL is that, contrary to regular computer programs which are sequential, its statements are inherently *concurrent* (parallel). For that reason, VHDL is usually referred to as a *code* rather than a program. In VHDL, only statements placed inside a PROCESS, FUNCTION, or PROCEDURE are executed sequentially.

1.2 Design Flow

As mentioned above, one of the major utilities of VHDL is that it allows the synthesis of a circuit or system in a programmable device (PLD or FPGA) or in an ASIC. The steps followed during such a project are summarized in figure 1.1. We start the design by writing the VHDL code, which is saved in a file with the extension

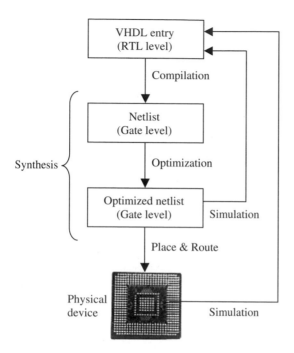

Figure 1.1
Summary of VHDL design flow.

.vhd and the same name as its ENTITY's name. The first step in the synthesis process is compilation. Compilation is the conversion of the high-level VHDL language, which describes the circuit at the Register Transfer Level (RTL), into a netlist at the gate level. The second step is optimization, which is performed on the gate-level netlist for speed or for area. At this stage, the design can be simulated. Finally, a place-and-route (fitter) software will generate the physical layout for a PLD/FPGA chip or will generate the masks for an ASIC.

1.3   EDA Tools

There are several EDA (Electronic Design Automation) tools available for circuit synthesis, implementation, and simulation using VHDL. Some tools (place and route, for example) are offered as part of a vendor's design suite (e.g., Altera's Quartus II, which allows the synthesis of VHDL code onto Altera's CPLD/FPGA chips, or Xilinx's ISE suite, for Xilinx's CPLD/FPGA chips). Other tools (synthesizers, for example), besides being offered as part of the design suites, can also be provided by specialized EDA companies (Mentor Graphics, Synopsis, Synplicity, etc.). Examples of the latter group are Leonardo Spectrum (a synthesizer from Mentor Graphics), Synplify (a synthesizer from Synplicity), and ModelSim (a simulator from Model Technology, a Mentor Graphics company).

The designs presented in the book were synthesized onto CPLD/FPGA devices (appendix A) either from Altera or Xilinx. The tools used were either ISE combined with ModelSim (for Xilinx chips—appendix B), MaxPlus II combined with Advanced Synthesis Software (for Altera CPLDs—appendix C), or Quartus II (also for Altera devices—appendix D). Leonardo Spectrum was also used occasionally.

Although different EDA tools were used to implement and test the examples presented in the book (see list of tools above), we decided to standardize the visual presentation of all simulation graphs. Due to its clean appearance, the waveform editor of MaxPlus II (appendix C) was employed. However, newer simulators, like ISE + ModelSim (appendix B) and Quartus II (appendix D), offer a much broader set of features, which allow, for example, a more refined timing analysis. For that reason, those tools were adopted when examining the fine details of each design.

1.4 Translation of VHDL Code into a Circuit

A full-adder unit is depicted in figure 1.2. In it, a and b represent the input bits to be added, cin is the carry-in bit, s is the sum bit, and cout the carry-out bit. As shown in the truth table, s must be high whenever the number of inputs that are high is odd, while cout must be high when two or more inputs are high.

A VHDL code for the full adder of figure 1.2 is shown in figure 1.3. As can be seen, it consists of an ENTITY, which is a description of the pins (PORTS) of the

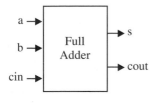

a b	cin	s	cout
0 0	0	0	0
0 1	0	1	0
1 0	0	1	0
1 1	0	0	1
0 0	1	1	0
0 1	1	0	1
1 0	1	0	1
1 1	1	1	1

Figure 1.2
Full-adder diagram and truth table.

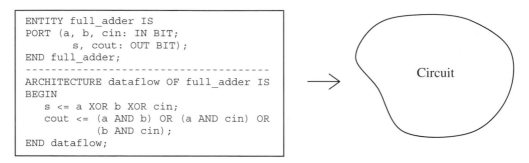

```
ENTITY full_adder IS
PORT (a, b, cin: IN BIT;
       s, cout: OUT BIT);
END full_adder;
----------------------------------------
ARCHITECTURE dataflow OF full_adder IS
BEGIN
    s <= a XOR b XOR cin;
    cout <= (a AND b) OR (a AND cin) OR
            (b AND cin);
END dataflow;
```

Figure 1.3
Example of VHDL code for the full-adder unit of figure 1.2.

circuit, and of an ARCHITECTURE, which describes how the circuit should function. We see in the latter that the sum bit is computed as $s = a \oplus b \oplus cin$, while cout is obtained from $cout = a.b + a.cin + b.cin$.

From the VHDL code shown on the left-hand side of figure 1.3, a physical circuit is inferred, as indicated on the right-hand side of the figure. However, there are several ways of implementing the equations described in the ARCHITECTURE of figure 1.3, so the actual circuit will depend on the compiler/optimizer being used and, more importantly, on the target technology. A few examples are presented in figure 1.4. For instance, if our target is a programmable logic device (PLD or FPGA—appendix A), then two possible results (among many others) for cout are illustrated in figures 1.4(b)–(c) (in both, of course, $cout = a.b + a.cin + b.cin$). On the other hand, if our target technology is an ASIC, then a possible CMOS implementation, at the transistor level, is that of figure 1.4(d) (which makes use of MOS transistors and clocked domino logic). Moreover, the synthesis tool can be set to optimize the layout for area or for speed, which obviously also affects the final circuitry.

Whatever the final circuit inferred from the code is, its operation should always be verified still at the design level (after synthesis), as indicated in figure 1.1. Of course, it must also be tested at the physical level, but then changes in the design might be too costly.

When testing, waveforms similar to those depicted in figure 1.5 will be displayed by the simulator. Indeed, figure 1.5 contains the simulation results from the circuit synthesized with the VHDL code of figure 1.3, which implements the full-adder unit of figure 1.2. As can be seen, the input pins (characterized by an inward arrow with an I marked inside) and the output pins (characterized by an outward arrow with an O marked inside) are those listed in the ENTITY of figure 1.3. We can freely estab-

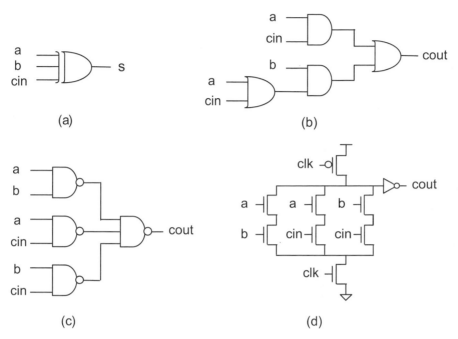

Figure 1.4
Examples of possible circuits obtained from the full-adder VHDL code of figure 1.3.

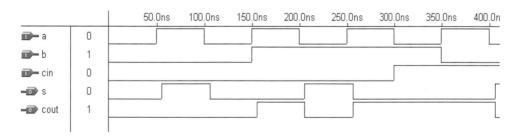

Figure 1.5
Simulation results from the VHDL design of figure 1.3.

lish the values of the input signals (a, b, and cin in this case), and the simulator will compute and plot the output signals (s and cout). As can be observed in figure 1.5, the outputs do behave as expected.

1.5   Design Examples

As mentioned in the preface, the book is indeed a *design-oriented* approach to the task of teaching VHDL. The integration between VHDL and Digital Design is achieved through a long series of well-detailed design examples. A summary of the *complete* designs presented in the book is shown below.

- Adders (examples 3.3 and 6.8 and section 9.3)
- ALU (examples 5.5 and 6.10)
- Barrel shifters and vector shifters (examples 5.6 and 6.9 and section 9.1)
- Comparators (section 9.2)
- Controller, traffic light (example 8.5)
- Controller, vending machine (section 9.5)
- Count ones (examples 7.1 and 7.2)
- Counters (examples 6.2, 6.5, 6.7, 7.7, and 8.1)
- Decoder (example 4.1)
- Digital filters (section 12.4)
- Dividers, fixed point (section 9.4)
- Flip-flops and latches (examples 2.1, 5.7, 5.8, 6.1, 6.4, 6.6, 7.4, and 7.6)
- Encoder (example 5.4)
- Frequency divider (example 7.5)
- Function arith_shift (example 11.7)
- Function conv_integer (examples 11.2 and 11.5)
- Function multiplier (example 11.8)
- Function "+" overloaded (example 11.6)
- Function positive_edge (examples 11.1, 11.3, and 11.4)
- Leading zeros counter (example 6.10)
- Multiplexers (examples 5.1, 5.2, and 7.3)

- Multipliers (example 11.8 and sections 12.1 and 12.2)
- MAC circuit (section 12.3)
- Neural networks (section 12.5)
- Parallel-to-serial converter (section 9.7)
- Parity detector (example 4.2)
- Parity generator (example 4.3)
- Playing with SSD (section 9.8)
- Procedure min_max (examples 11.9 and 11.10)
- RAM (example 6.11 and section 9.10)
- ROM (section 9.10)
- Serial data receiver (section 9.6)
- Shift registers (examples 6.3, 7.8, and 7.9)
- Signal generators (example 8.6 and section 9.9)
- String detector (example 8.4)
- Tri-state buffer/bus (example 5.3)

Moreover, several additional designs and experimental verifications are also proposed as exercises:

- Adders and subtractors (problems 3.5, 5.4, 5.5, 6.14, 6.16, 10.2, and 10.3)
- Arithmetic-logic units (problems 6.13 and 10.1)
- Barrel and vector shifters (problems 5.7, 6.12, 9.1, and 12.2)
- Binary-to-Gray code converter (problem 5.6)
- Comparators (problems 5.8 and 6.15)
- Count ones (problem 6.9)
- Counters (problems 7.5 and 11.6)
- Data delay circuit (problem 7.2)
- Decoders (problems 4.4 and 7.6)
- DFFs (problems 6.17, 7.3, 7.4, and 7.7)
- Digital FIR filter (problem 12.4)
- Dividers (problems 5.3 and 9.2)
- Event counter (problem 6.1)

- Finite-state machine (problem 8.1)
- Frequency divider, generic (problem 6.4)
- Frequency multiplier (problem 6.5)
- Function conv_std_logic_vector (problem 11.1)
- Function "not" overloaded for integers (problem 11.2)
- Function shift for integers (problem 11.4)
- Function shift for std_logic_vector (problem 11.3)
- Function BCD-SSD converter (problem 11.6)
- Function "+" overloaded for std_logic_vector (problem 11.8)
- Intensity encoder (problem 6.10)
- Keypad debouncer/encoder (problem 8.4)
- Multiplexers (problems 2.1, 5.1, and 6.11)
- Multipliers (problems 5.3, 11.5, and 12.1)
- Multiply-accumulate circuit (problem 12.3)
- Neural network (problem 12.5)
- Parity detector (problem 6.8)
- Playing with a seven-segment display (problem 9.6)
- Priority encoder (problems 5.2 and 6.3)
- Procedure statistics (problem 11.7)
- Random number generator plus SSD (problem 9.8)
- ROM (problem 3.4)
- Serial data receiver (problem 9.4)
- Serial data transmitter (problem 9.5)
- Shift register (problem 6.2)
- Signal generators (problems 8.2, 8.3, 8.6, and 8.7)
- Speed monitor (problem 9.7)
- Stop watch (problem 10.4)
- Timers (problems 6.6 and 6.7)
- Traffic-light controller (problem 8.5)
- Vending-machine controller (problem 9.3)

Additionally, four appendices on programmable logic devices and synthesis tools are included:

· Appendix A: Programmable Logic Devices
· Appendix B: Xilinx ISE + ModelSim Tutorial
· Appendix C: Altera MaxPlus II + Advanced Synthesis Software Tutorial
· Appendix D: Altera Quartus II Tutorial

2 Code Structure

In this chapter, we describe the fundamental sections that comprise a piece of VHDL code: LIBRARY declarations, ENTITY, and ARCHITECTURE.

2.1 Fundamental VHDL Units

As depicted in figure 2.1, a standalone piece of VHDL code is composed of at least three fundamental sections:

• LIBRARY declarations: Contains a list of all libraries to be used in the design. For example: *ieee*, *std*, *work*, etc.

• ENTITY: Specifies the I/O pins of the circuit.

• ARCHITECTURE: Contains the VHDL code proper, which describes how the circuit should behave (function).

A LIBRARY is a collection of commonly used pieces of code. Placing such pieces inside a library allows them to be reused or shared by other designs.

The typical structure of a library is illustrated in figure 2.2. The code is usually written in the form of FUNCTIONS, PROCEDURES, or COMPONENTS, which are placed inside PACKAGES, and then compiled into the destination library.

The fundamental units of VHDL (figure 2.1) will be studied in Part I of the book (up to chapter 9), whereas the library-related sections (figure 2.2) will be seen in Part II (chapters 10–12).

2.2 LIBRARY Declarations

To declare a LIBRARY (that is, to make it visible to the design) two lines of code are needed, one containing the *name* of the library, and the other a *use* clause, as shown in the syntax below.

```
LIBRARY library_name;
USE library_name.package_name.package_parts;
```

At least three packages, from three different libraries, are usually needed in a design:

• *ieee.std_logic_1164* (from the *ieee* library),

• *standard* (from the *std* library), and

• *work* (*work* library).

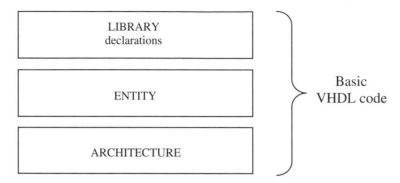

Figure 2.1
Fundamental sections of a basic VHDL code.

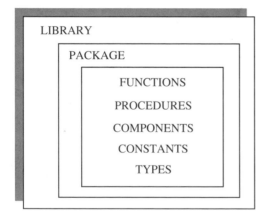

Figure 2.2
Fundamental parts of a LIBRARY.

Their declarations are as follows:

```
LIBRARY ieee;              -- A semi-colon (;) indicates
USE ieee.std_logic_1164.all;   -- the end of a statement or

LIBRARY std;               -- declaration, while a double
USE std.standard.all;      -- dash (--) indicates a comment.

LIBRARY work;
USE work.all;
```

The libraries *std* and *work* shown above are made visible by default, so there is no need to declare them; only the *ieee* library must be explicitly written. However, the latter is only necessary when the STD_LOGIC (or STD_ULOGIC) data type is employed in the design (data types will be studied in detail in the next chapter).

The purpose of the three packages/libraries mentioned above is the following: the *std_logic_1164* package of the *ieee* library specifies a multi-level logic system; *std* is a resource library (data types, text i/o, etc.) for the VHDL design environment; and the *work* library is where we save our design (the .vhd file, plus all files created by the compiler, simulator, etc.).

Indeed, the *ieee* library contains several packages, including the following:

• *std_logic_1164*: Specifies the STD_LOGIC (8 levels) and STD_ULOGIC (9 levels) multi-valued logic systems.

• *std_logic_arith*: Specifies the SIGNED and UNSIGNED data types and related arithmetic and comparison operations. It also contains several data conversion functions, which allow one type to be converted into another: *conv_integer(p)*, *conv_unsigned(p, b)*, *conv_signed(p, b)*, *conv_std_logic_vector(p, b)*.

• *std_logic_signed*: Contains functions that allow operations with STD_LOGIC_VECTOR data to be performed as if the data were of type SIGNED.

• *std_logic_unsigned*: Contains functions that allow operations with STD_LOGIC_VECTOR data to be performed as if the data were of type UNSIGNED.

In chapter 3, all these libraries will be further described and used.

2.3 ENTITY

An ENTITY is a list with specifications of all input and output pins (PORTS) of the circuit. Its syntax is shown below.

```
ENTITY entity_name IS
   PORT (
       port_name : signal_mode signal_type;
       port_name : signal_mode signal_type;
       ...);
END entity_name;
```

The *mode* of the signal can be IN, OUT, INOUT, or BUFFER. As illustrated in figure 2.3, IN and OUT are truly unidirectional pins, while INOUT is bidirectional. BUFFER, on the other hand, is employed when the output signal must be used (read) internally.

The *type* of the signal can be BIT, STD_LOGIC, INTEGER, etc. Data types will be discussed in detail in chapter 3.

Finally, the *name* of the entity can be basically any name, except VHDL reserved words (VHDL reserved words are listed in appendix E).

Example: Let us consider the NAND gate of figure 2.4. Its ENTITY can be specified as:

```
ENTITY nand_gate IS
   PORT (a, b : IN BIT;
         x : OUT BIT);
END nand_gate;
```

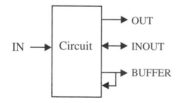

Figure 2.3
Signal *modes.*

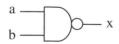

Figure 2.4
NAND gate.

The meaning of the ENTITY above is the following: the circuit has three I/O pins, being two inputs (a and b, mode IN) and one output (x, mode OUT). All three signals are of type BIT. The name chosen for the entity was nand_gate.

2.4   ARCHITECTURE

The ARCHITECTURE is a description of how the circuit should behave (function). Its syntax is the following:

```
ARCHITECTURE architecture_name OF entity_name IS
    [declarations]
BEGIN
    (code)
END architecture_name;
```

As shown above, an architecture has two parts: a *declarative* part (optional), where signals and constants (among others) are declared, and the *code* part (from BEGIN down). Like in the case of an entity, the name of an architecture can be basically any name (except VHDL reserved words), including the same name as the entity's.

Example: Let us consider the NAND gate of figure 2.4 once again.

```
ARCHITECTURE myarch OF nand_gate IS
BEGIN
    x <= a NAND b;
END myarch;
```

The meaning of the ARCHITECTURE above is the following: the circuit must perform the NAND operation between the two input signals (a, b) and assign ("<=") the result to the output pin (x). The name chosen for this architecture was myarch. In this example, there is no declarative part, and the code contains just a single assignment.

2.5   Introductory Examples

In this section, we will present two initial examples of VHDL code. Though we have not yet studied the constructs that appear in the examples, they will help illustrate fundamental aspects regarding the overall code structure. Each example is followed by explanatory comments and simulation results.

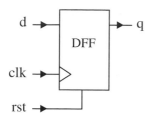

Figure 2.5
DFF with asynchronous reset.

Example 2.1:   DFF with Asynchronous Reset

Figure 2.5 shows the diagram of a D-type flip-flop (DFF), triggered at the rising-edge of the clock signal (clk), and with an asynchronous reset input (rst). When rst = '1', the output must be turned low, regardless of clk. Otherwise, the output must copy the input (that is, q <= d) at the moment when clk changes from '0' to '1' (that is, when an upward event occurs on clk).

There are several ways of implementing the DFF of figure 2.5, one being the solution presented below. One thing to remember, however, is that VHDL is inherently concurrent (contrary to regular computer programs, which are sequential), so to implement any clocked circuit (flip-flops, for example) we have to "force" VHDL to be sequential. This can be done using a PROCESS, as shown below.

```
1   ----------------------------------------
2   LIBRARY ieee;
3   USE ieee.std_logic_1164.all;
4   ----------------------------------------
5   ENTITY dff IS
6      PORT ( d, clk, rst: IN STD_LOGIC;
7              q: OUT STD_LOGIC);
8   END dff;
9   ----------------------------------------
10  ARCHITECTURE behavior OF dff IS
11  BEGIN
12     PROCESS (rst, clk)
13     BEGIN
14        IF (rst='1') THEN
15           q <= '0';
16        ELSIF (clk'EVENT AND clk='1') THEN
```

```
17          q <= d;
18       END IF;
19    END PROCESS;
20 END behavior;
21 -------------------------------------
```

Comments:

Lines 2–3: Library declaration (library *name* and library *use* clause). Recall that the other two indispensable libraries (*std* and *work*) are made visible by default.

Lines 5–8: Entity *dff*.

Lines 10–20: Architecture *behavior*.

Line 6: Input ports (input mode can only be IN). In this example, all input signals are of type STD_LOGIC.

Line 7: Output port (output mode can be OUT, INOUT, or BUFFER). Here, the output is also of type STD_LOGIC.

Lines 11–19: Code part of the architecture (from word BEGIN on).

Lines 12–19: A PROCESS (inside it the code is executed sequentially).

Line 12: The PROCESS is executed every time a signal declared in its sensitivity list changes. In this example, every time rst or clk changes the PROCESS is run.

Lines 14–15: Every time rst goes to '1' the output is reset, regardless of clk (asynchronous reset).

Lines 16–17: If rst is not active, plus clk has changed (an EVENT occurred on clk), plus such event was a rising edge (clk = '1'), then the input signal (d) is stored in the flip-flop (q <= d).

Lines 15 and 17: The "<=" operator is used to assign a value to a SIGNAL. In contrast, ":=" would be used for a VARIABLE. All ports in an entity are signals by default.

Lines 1, 4, 9, and 21: Commented out (recall that "--" indicates a comment). Used only to better organize the design.

Note: VHDL is *not* case sensitive.

Simulation results:
Figure 2.6 presents simulation results regarding example 2.1. The graphs can be easily interpreted. The first column shows the signal names, as defined in the ENTITY. It also shows the mode (direction) of the signals; notice that the arrows associated

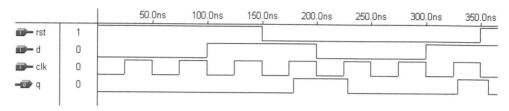

Figure 2.6
Simulation results of example 2.1.

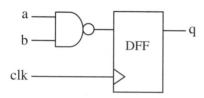

Figure 2.7
DFF plus NAND gate.

with rst, d, and clk are inward, and contain the letter I (input) inside, while that of q
is outward and has an O (output) marked inside. The second column has the value of
each signal in the position where the vertical cursor is placed. In the present case, the
cursor is at 0ns, where the signals have value 1, 0, 0, 0, respectively. In this example,
the values are simply '0' or '1', but when vectors are used, the values can be shown in
binary, decimal, or hexadecimal form. The third column shows the simulation
proper. The input signals (rst, d, clk) can be chosen freely, and the simulator will
determine the corresponding output (q). Comparing the results of figure 2.6 with
those expected from the circuit shown previously, we notice that it works properly.
As mentioned earlier, the designs presented in the book were synthesized onto CPLD/
FPGA devices (appendix A), either from Altera or Xilinx. The tools used were either
ISE combined with ModelSim (for Xilinx chips—appendix B), or MaxPlus II com-
bined with Advanced Synthesis Software (for Altera CPLDs—appendix C), or
Quartus II (also for Altera devices—appendix D). Leonardo Spectrum (from Mentor
Graphics) was also used occasionally.

Example 2.2: DFF plus NAND Gate

The circuit of figure 2.4 was purely combinational, while that of figure 2.5 was purely
sequential. The circuit of figure 2.7 is a mixture of both (without reset). In the

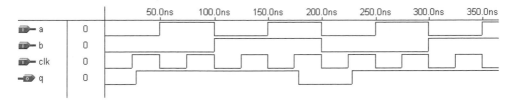

Figure 2.8
Simulation results of example 2.2.

solution that follows, we have purposely introduced an unnecessary signal (temp), just to illustrate how a signal should be declared. Simulation results from the circuit synthesized with the code below are shown in figure 2.8.

```
1   ----------------------------------------
2   ENTITY example IS
3      PORT ( a, b, clk: IN BIT;
4               q: OUT BIT);
5   END example;
6   ----------------------------------------
7   ARCHITECTURE example OF example IS
8      SIGNAL temp : BIT;
9   BEGIN
10     temp <= a NAND b;
11     PROCESS (clk)
12     BEGIN
13        IF (clk'EVENT AND clk='1') THEN q<=temp;
14        END IF;
15     END PROCESS;
16  END example;
17  ----------------------------------------
```

Comments:
Library declarations are not necessary in this case, because the data is of type BIT, which is specified in the library *std* (recall that the libraries *std* and *work* are made visible by default).

Lines 2–5: Entity *example*.

Lines 7–16: Architecture *example*.

Line 3: Input ports (all of type BIT).

Line 4: Output port (also of type BIT).

Line 8: Declarative part of the architecture (optional). The signal *temp*, of type BIT, was declared. Notice that there is no *mode* declaration (*mode* is only used in entities).

Lines 9–15: Code part of the architecture (from word BEGIN on).

Lines 11–15: A PROCESS (sequential statements executed every time the signal clk changes).

Lines 10 and 11–15: Though within a process the execution is sequential, the process, as a whole, is concurrent with the other (external) statements; thus line 10 is executed concurrently with the block 11–15.

Line 10: Logical NAND operation. Result is assigned to signal *temp*.

Lines 13–14: IF statement. At the rising edge of clk the value of temp is assigned to q.

Lines 10 and 13: The "$<=$" operator is used to assign a value to a SIGNAL. In contrast, "$:=$" would be used for a VARIABLE.

Lines 8 and 10: Can be eliminated, changing "q $<=$ a NAND b" in line 13.

Lines 1, 6, and 17: Commented out. Used only to better organize the design.

2.6   Problems

Problem 2.1:   Multiplexer

The top-level diagram of a multiplexer is shown in figure P2.1. According to the truth table, the output should be equal to one of the inputs if sel $=$ "01" (c $=$ a) or sel $=$ "10" (c $=$ b), but it should be zero or high impedance if sel $=$ "00" or sel $=$ "11", respectively.

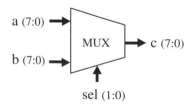

sel	c
00	00000000
01	a
10	b
11	ZZZZZZZZ

Figure P2.1

a) Complete the VHDL code below.

b) Write relevant comments regarding your solution (as in examples 2.1 and 2.2).

c) Compile and simulate your solution, checking whether it works as expected.

Note: A solution using IF was employed in the code below, because it is more intuitive. However, as will be seen later, a multiplexer can also be implemented with other statements, like WHEN or CASE.

```
1  ----------------------------------------
2  LIBRARY ieee;
3  USE _____ ;
4  ----------------------------------------
5  ENTITY mux IS
6      PORT ( __ , __ : ___ STD_LOGIC_VECTOR (7 DOWNTO 0);
7             sel : IN _____ ;
8             ___ : OUT STD_LOGIC_VECTOR (7 DOWNTO 0));
9  END _____ ;
10 ----------------------------------------
11 ARCHITECTURE example OF _____ IS
12 BEGIN
13     PROCESS (a, b, ____ )
14     BEGIN
15        IF (sel = "00") THEN
16           c <= "00000000";
17        ELSIF (_____) THEN
18           c <= a;
19        _____ (sel = "10") THEN
20           c <= __;
21        ELSE
22           c <= (OTHERS => '__');
23        END ___ ;
24     END _____ ;
25 END _____ ;
26 ----------------------------------------
```

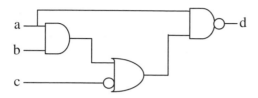

Figure P2.2

Problem 2.2:   Logic Gates

a) Write a VHDL code for the circuit of figure P2.2. Notice that it is purely combinational, so a PROCESS is not necessary. Write an expression for d using only logical operators (AND, OR, NAND, NOT, etc.).

b) Synthesize and simulate your circuit. After assuring that it works properly, open the report file and check the actual expression implemented by the compiler. Compare it with your expression.

3 Data Types

In order to write VHDL code efficiently, it is essential to know what *data types* are allowed, and how to specify and use them. In this chapter, all fundamental data types are described, with special emphasis on those that are synthesizable. Discussions on data compatibility and data conversion are also included.

3.1 Pre-Defined Data Types

VHDL contains a series of pre-defined data types, specified through the IEEE 1076 and IEEE 1164 standards. More specifically, such data type definitions can be found in the following packages / libraries:

• Package *standard* of library *std*: Defines BIT, BOOLEAN, INTEGER, and REAL data types.

• Package *std_logic_1164* of library *ieee*: Defines STD_LOGIC and STD_ULOGIC data types.

• Package *std_logic_arith* of library *ieee*: Defines SIGNED and UNSIGNED data types, plus several data conversion functions, like *conv_integer(p)*, *conv_unsigned(p, b)*, *conv_signed(p, b)*, and *conv_std_logic_vector(p, b)*.

• Packages *std_logic_signed* and *std_logic_unsigned* of library *ieee*: Contain functions that allow operations with STD_LOGIC_VECTOR data to be performed as if the data were of type SIGNED or UNSIGNED, respectively.

All pre-defined data types (specified in the packages/libraries listed above) are described below.

• BIT (and BIT_VECTOR): 2-level logic ('0', '1').

Examples:

```
SIGNAL x: BIT;
-- x is declared as a one-digit signal of type BIT.

SIGNAL y: BIT_VECTOR (3 DOWNTO 0);
-- y is a 4-bit vector, with the leftmost bit being the MSB.

SIGNAL w: BIT_VECTOR (0 TO 7);
-- w is an 8-bit vector, with the rightmost bit being the MSB.
```

Based on the signals above, the following assignments would be legal (to assign a value to a signal, the "<=" operator must be used):

```
x <= '1';
-- x is a single-bit signal (as specified above), whose value is
-- '1'. Notice that single quotes (' ') are used for a single bit.

y <= "0111";
-- y is a 4-bit signal (as specified above), whose value is "0111"
-- (MSB='0'). Notice that double quotes (" ") are used for
-- vectors.

w <= "01110001";
-- w is an 8-bit signal, whose value is "01110001" (MSB='1').
```

• STD_LOGIC (and STD_LOGIC_VECTOR): 8-valued logic system introduced in the IEEE 1164 standard.

'X'	Forcing Unknown	(synthesizable unknown, with restrictions)
'0'	Forcing Low	(synthesizable logic '0')
'1'	Forcing High	(synthesizable logic '1')
'Z'	High impedance	(synthesizable tri-state buffer)
'W'	Weak unknown	
'L'	Weak low	
'H'	Weak high	
'–'	Don't care	

Examples:

```
SIGNAL x: STD_LOGIC;
-- x is declared as a one-digit (scalar) signal of type STD_LOGIC.

SIGNAL y: STD_LOGIC_VECTOR (3 DOWNTO 0) := "0001";
-- y is declared as a 4-bit vector, with the leftmost bit being
-- the MSB. The initial value (optional) of y is "0001". Notice
-- that the ":=" operator is used to establish the initial value.
```

Most of the std_logic levels are intended for simulation only. However, '0', '1', and 'Z' are synthesizable with no restrictions. With respect to the "weak" values, they are resolved in favor of the "forcing" values in multiply-driven nodes (see table 3.1). Indeed, if any two std_logic signals are connected to the same node, then conflicting logic levels are automatically resolved according to table 3.1.

• STD_ULOGIC (STD_ULOGIC_VECTOR): 9-level logic system introduced in the IEEE 1164 standard ('U', 'X', '0', '1', 'Z', 'W', 'L', 'H', '–'). Indeed, the

Table 3.1
Resolved logic system (STD_LOGIC).

	X	0	1	Z	W	L	H	-
X	X	X	X	X	X	X	X	X
0	X	0	X	0	0	0	0	X
1	X	X	1	1	1	1	1	X
Z	X	0	1	Z	W	L	H	X
W	X	0	1	W	W	W	W	X
L	X	0	1	L	W	L	W	X
H	X	0	1	H	W	W	H	X
-	X	X	X	X	X	X	X	X

STD_LOGIC system described above is a *subtype* of STD_ULOGIC. The latter includes an extra logic value, 'U', which stands for unresolved. Thus, contrary to STD_LOGIC, conflicting logic levels are not automatically resolved here, so output wires should never be connected together directly. However, if two output wires are never supposed to be connected together, this logic system can be used to detect design errors.

• BOOLEAN: True, False.

• INTEGER: 32-bit integers (from −2,147,483,647 to +2,147,483,647).

• NATURAL: Non-negative integers (from 0 to +2,147,483,647).

• REAL: Real numbers ranging from −1.0E38 to +1.0E38. Not synthesizable.

• Physical literals: Used to inform physical quantities, like time, voltage, etc. Useful in simulations. Not synthesizable.

• Character literals: Single ASCII character or a string of such characters. Not synthesizable.

• SIGNED and UNSIGNED: data types defined in the *std_logic_arith* package of the *ieee* library. They have the appearance of STD_LOGIC_VECTOR, but accept arithmetic operations, which are typical of INTEGER data types (SIGNED and UNSIGNED will be discussed in detail in section 3.6).

Examples:

```
x0 <= '0';               -- bit, std_logic, or std_ulogic value '0'
x1 <= "00011111";        -- bit_vector, std_logic_vector,
                         -- std_ulogic_vector, signed, or unsigned
x2 <= "0001_1111";       -- underscore allowed to ease visualization
x3 <= "101111"           -- binary representation of decimal 47
```

```
x4 <= B"101111"        -- binary representation of decimal 47
x5 <= O"57"            -- octal representation of decimal 47
x6 <= X"2F"            -- hexadecimal representation of decimal 47
n <= 1200;             -- integer
m <= 1_200;            -- integer, underscore allowed
IF ready THEN...       -- Boolean, executed if ready=TRUE
y <= 1.2E-5;           -- real, not synthesizable
q <= d after 10 ns;    -- physical, not synthesizable
```

Example: Legal and illegal operations between data of different types.

```
SIGNAL a: BIT;
SIGNAL b: BIT_VECTOR(7 DOWNTO 0);
SIGNAL c: STD_LOGIC;
SIGNAL d: STD_LOGIC_VECTOR(7 DOWNTO 0);
SIGNAL e: INTEGER RANGE 0 TO 255;
...
a <= b(5);    -- legal (same scalar type: BIT)
b(0) <= a;    -- legal (same scalar type: BIT)
c <= d(5);    -- legal (same scalar type: STD_LOGIC)
d(0) <= c;    -- legal (same scalar type: STD_LOGIC)
a <= c;       -- illegal (type mismatch: BIT x STD_LOGIC)
b <= d;       -- illegal (type mismatch: BIT_VECTOR x
              -- STD_LOGIC_VECTOR)
e <= b;       -- illegal (type mismatch: INTEGER x BIT_VECTOR)
e <= d;       -- illegal (type mismatch: INTEGER x
              -- STD_LOGIC_VECTOR)
```

3.2 User-Defined Data Types

VHDL also allows the user to define his/her own data types. Two categories of user-defined data types are shown below: *integer* and *enumerated*.

• User-defined *integer* types:

```
TYPE integer IS RANGE -2147483647 TO +2147483647;
-- This is indeed the pre-defined type INTEGER.

TYPE natural IS RANGE 0 TO +2147483647;
-- This is indeed the pre-defined type NATURAL.
```

```
TYPE my_integer IS RANGE -32 TO 32;
-- A user-defined subset of integers.

TYPE student_grade IS RANGE 0 TO 100;
-- A user-defined subset of integers or naturals.
```

• User-defined *enumerated* types:

```
TYPE bit IS ('0', '1');
-- This is indeed the pre-defined type BIT

TYPE my_logic IS ('0', '1', 'Z');
-- A user-defined subset of std_logic.

TYPE bit_vector IS ARRAY (NATURAL RANGE <>) OF BIT;
-- This is indeed the pre-defined type BIT_VECTOR.
-- RANGE <> is used to indicate that the range is unconstrained.
-- NATURAL RANGE <>, on the other hand, indicates that the only
-- restriction is that the range must fall within the NATURAL
-- range.

TYPE state IS (idle, forward, backward, stop);
-- An enumerated data type, typical of finite state machines.

TYPE color IS (red, green, blue, white);
-- Another enumerated data type.
```

The encoding of enumerated types is done sequentially and automatically (unless specified otherwise by a user-defined attribute, as will be shown in chapter 4). For example, for the type *color* above, two bits are necessary (there are four states), being "00" assigned to the first state (red), "01" to the second (green), "10" to the next (blue), and finally "11" to the last state (white).

3.3 Subtypes

A SUBTYPE is a TYPE with a constraint. The main reason for using a subtype rather than specifying a new type is that, though operations between data of different types are not allowed, they are allowed between a subtype and its corresponding base type.

Examples: The subtypes below were derived from the types presented in the previous examples.

```
SUBTYPE natural IS INTEGER RANGE 0 TO INTEGER'HIGH;
-- As expected, NATURAL is a subtype (subset) of INTEGER.

SUBTYPE my_logic IS STD_LOGIC RANGE '0' TO 'Z';
-- Recall that STD_LOGIC=('X','0','1','Z','W','L','H','-').
-- Therefore, my_logic=('0','1','Z').

SUBTYPE my_color IS color RANGE red TO blue;
-- Since color=(red, green, blue, white), then
-- my_color=(red, green, blue).

SUBTYPE small_integer IS INTEGER RANGE -32 TO 32;
-- A subtype of INTEGER.
```

Example: Legal and illegal operations between types and subtypes.

```
SUBTYPE my_logic IS STD_LOGIC RANGE '0' TO '1';
SIGNAL a: BIT;
SIGNAL b: STD_LOGIC;
SIGNAL c: my_logic;
...
b <= a;    -- illegal (type mismatch: BIT versus STD_LOGIC)
b <= c;    -- legal (same "base" type: STD_LOGIC)
```

3.4 Arrays

Arrays are collections of objects of the same type. They can be one-dimensional (1D), two-dimensional (2D), or one-dimensional-by-one-dimensional (1Dx1D). They can also be of higher dimensions, but then they are generally not synthesizable.

Figure 3.1 illustrates the construction of data arrays. A single value (scalar) is shown in (a), a vector (1D array) in (b), an array of vectors (1Dx1D array) in (c), and an array of scalars (2D array) in (d).

Indeed, the pre-defined VHDL data types (seen in section 3.1) include only the scalar (single bit) and vector (one-dimensional array of bits) categories. The pre-defined *synthesizable* types in each of these categories are the following:

• Scalars: BIT, STD_LOGIC, STD_ULOGIC, and BOOLEAN.

• Vectors: BIT_VECTOR, STD_LOGIC_VECTOR, STD_ULOGIC_VECTOR, INTEGER, SIGNED, and UNSIGNED.

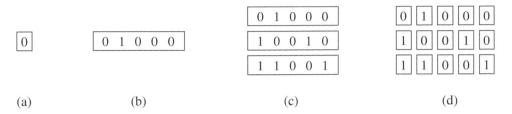

Figure 3.1
Illustration of (a) scalar, (b) 1D, (c) 1Dx1D, and (d) 2D data arrays.

As can be seen, there are no pre-defined 2D or 1Dx1D arrays, which, when necessary, must be specified by the user. To do so, the new TYPE must first be defined, then the new SIGNAL, VARIABLE, or CONSTANT can be declared using that data type. The syntax below should be used.

To specify a new array type:

```
TYPE type_name IS ARRAY (specification) OF data_type;
```

To make use of the new array type:

```
SIGNAL signal_name: type_name [:= initial_value];
```

In the syntax above, a SIGNAL was declared. However, it could also be a CONSTANT or a VARIABLE. Notice that the initial value is optional (for simulation only).

Example: 1Dx1D array.
Say that we want to build an array containing four vectors, each of size eight bits. This is then an 1Dx1D array (see figure 3.1). Let us call each vector by *row*, and the complete array by *matrix*. Additionally, say that we want the leftmost bit of each vector to be its MSB (most significant bit), and that we want the top row to be row 0. Then the array implementation would be the following (notice that a signal, called x, of type matrix, was declared as an example):

```
TYPE row IS ARRAY (7 DOWNTO 0) OF STD_LOGIC;     -- 1D array
TYPE matrix IS ARRAY (0 TO 3) OF row;            -- 1Dx1D array
SIGNAL x: matrix;                                -- 1Dx1D signal
```

Example: Another 1Dx1D array.

Another way of constructing the 1Dx1D array above would be the following:

```
TYPE matrix IS ARRAY (0 TO 3) OF STD_LOGIC_VECTOR(7 DOWNTO 0);
```

   From a data-compatibility point of view, the latter might be advantageous over that in the previous example (see example 3.1).

Example: 2D array.

The array below is truly two-dimensional. Notice that its construction is not based on vectors, but rather entirely on scalars.

```
TYPE matrix2D IS ARRAY (0 TO 3, 7 DOWNTO 0) OF STD_LOGIC;
-- 2D array
```

Example: Array initialization.

As shown in the syntax above, the initial value of a SIGNAL or VARIABLE is optional. However, when initialization is required, it can be done as in the examples below.

```
... :="0001";                              -- for 1D array
... :=('0','0','0','1')                    -- for 1D array
... :=(('0','1','1','1'), ('1','1','1','0'));  -- for 1Dx1D or
                                           -- 2D array
```

Example: Legal and illegal array assignments.

The assignments in this example are based on the following type definitions and signal declarations:

```
TYPE row IS ARRAY (7 DOWNTO 0) OF STD_LOGIC;
                                           -- 1D array
TYPE array1 IS ARRAY (0 TO 3) OF row;
                                           -- 1Dx1D array
TYPE array2 IS ARRAY (0 TO 3) OF STD_LOGIC_VECTOR(7 DOWNTO 0);
                                           -- 1Dx1D
TYPE array3 IS ARRAY (0 TO 3, 7 DOWNTO 0) OF STD_LOGIC;
                                           -- 2D array
SIGNAL x: row;
SIGNAL y: array1;
SIGNAL v: array2;
SIGNAL w: array3;
```

```
--------- Legal scalar assignments: --------------
-- The scalar (single bit) assignments below are all legal,
-- because the "base" (scalar) type is STD_LOGIC for all signals
-- (x,y,v,w).
x(0) <= y(1)(2);        -- notice two pairs of parenthesis
                        -- (y is 1Dx1D)
x(1) <= v(2)(3);        -- two pairs of parenthesis (v is 1Dx1D)
x(2) <= w(2,1);         -- a single pair of parenthesis (w is 2D)
y(1)(1) <= x(6);
y(2)(0) <= v(0)(0);
y(0)(0) <= w(3,3);
w(1,1) <= x(7);
w(3,0) <= v(0)(3);
--------- Vector assignments: --------------------
x <= y(0);                    -- legal (same data types: ROW)
x <= v(1);                    -- illegal (type mismatch: ROW x
                              -- STD_LOGIC_VECTOR)
x <= w(2);                    -- illegal (w must have 2D index)
x <= w(2, 2 DOWNTO 0);        -- illegal (type mismatch: ROW x
                              -- STD_LOGIC)
v(0) <= w(2, 2 DOWNTO 0);     -- illegal (mismatch: STD_LOGIC_VECTOR
                              -- x STD_LOGIC)
v(0) <= w(2);                 -- illegal (w must have 2D index)
y(1) <= v(3);                 -- illegal (type mismatch: ROW x
                              -- STD_LOGIC_VECTOR)
y(1)(7 DOWNTO 3) <= x(4 DOWNTO 0);    -- legal (same type,
                                      -- same size)
v(1)(7 DOWNTO 3) <= v(2)(4 DOWNTO 0);  -- legal (same type,
                                       -- same size)
w(1, 5 DOWNTO 1) <= v(2)(4 DOWNTO 0);  -- illegal (type mismatch)
```

3.5 Port Array

As we have seen, there are no pre-defined data types of more than one dimension.
However, in the specification of the input or output pins (PORTS) of a circuit (which
is made in the ENTITY), we might need to specify the ports as arrays of vectors.
Since TYPE declarations are not allowed in an ENTITY, the solution is to declare

user-defined data types in a PACKAGE, which will then be visible to the whole design (thus including the ENTITY). An example is shown below.

```
------- Package: --------------------------
LIBRARY ieee;
USE ieee.std_logic_1164.all;
----------------------------
PACKAGE my_data_types IS
   TYPE vector_array IS ARRAY (NATURAL RANGE <>) OF
      STD_LOGIC_VECTOR(7 DOWNTO 0);
END my_data_types;
-----------------------------------------

------- Main code: -----------------------
LIBRARY ieee;
USE ieee.std_logic_1164.all;
USE work.my_data_types.all;      -- user-defined package
---------------------------
ENTITY mux IS
   PORT (inp: IN VECTOR_ARRAY (0 TO 3);
   ... );
END mux;
   ... ;
-----------------------------------------
```

As can be seen in the example above, a user-defined data type, called *vector_array*, was created, which can contain an indefinite number of vectors of size eight bits each (NATURAL RANGE <> signifies that the range is not fixed, with the only restriction that it must fall within the NATURAL range, which goes from 0 to +2,147,483,647). The data type was saved in a PACKAGE called *my_data_types*, and later used in an ENTITY to specify a PORT called *inp*. Notice in the main code the inclusion of an additional USE clause to make the user-defined package *my_data_types* visible to the design.

Another option for the PACKAGE above would be that shown below, where a CONSTANT declaration is included (a detailed study of PACKAGES will be presented in chapter 10).

```
------- Package: -----------------------------
LIBRARY ieee;
USE ieee.std_logic_1164.all;
```

```
---------------------------
PACKAGE my_data_types IS
   CONSTANT b: INTEGER := 7;
   TYPE vector_array IS ARRAY (NATURAL RANGE <>) OF
      STD_LOGIC_VECTOR(b DOWNTO 0);
END my_data_types;
--------------------------------------------------
```

3.6 Records

Records are similar to arrays, with the only difference that they contain objects of *different* types.

Example:

```
TYPE birthday IS RECORD
   day: INTEGER RANGE 1 TO 31;
   month: month_name;
END RECORD;
```

3.7 Signed and Unsigned Data Types

As mentioned earlier, these types are defined in the *std_logic_arith* package of the *ieee* library. Their syntax is illustrated in the examples below.

Examples:

```
SIGNAL x: SIGNED (7 DOWNTO 0);
SIGNAL y: UNSIGNED (0 TO 3);
```

Notice that their syntax is similar to that of STD_LOGIC_VECTOR, not like that of an INTEGER, as one might have expected.

An UNSIGNED value is a number never lower than zero. For example, "0101" represents the decimal 5, while "1101" signifies 13. If type SIGNED is used instead, the value can be positive or negative (in two's complement format). Therefore, "0101" would represent the decimal 5, while "1101" would mean -3.

To use SIGNED or UNSIGNED data types, the *std_logic_arith* package, of the *ieee* library, must be declared. Despite their syntax, SIGNED and UNSIGNED data types are intended mainly for *arithmetic* operations, that is, contrary to

STD_LOGIC_VECTOR, they accept arithmetic operations. On the other hand, logical operations are not allowed. With respect to relational (comparison) operations, there are no restrictions.

Example: Legal and illegal operations with signed/unsigned data types.

```
LIBRARY ieee;
USE ieee.std_logic_1164.all;
USE ieee.std_logic_arith.all;    -- extra package necessary
...
SIGNAL a: IN SIGNED (7 DOWNTO 0);
SIGNAL b: IN SIGNED (7 DOWNTO 0);
SIGNAL x: OUT SIGNED (7 DOWNTO 0);
...
v <= a + b;      -- legal (arithmetic operation OK)
w <= a AND b;    -- illegal (logical operation not OK)
```

Example: Legal and illegal operations with std_logic_vector.

```
LIBRARY ieee;
USE ieee.std_logic_1164.all;     -- no extra package required
...
SIGNAL a: IN STD_LOGIC_VECTOR (7 DOWNTO 0);
SIGNAL b: IN STD_LOGIC_VECTOR (7 DOWNTO 0);
SIGNAL x: OUT STD_LOGIC_VECTOR (7 DOWNTO 0);
...
v <= a + b;      -- illegal (arithmetic operation not OK)
w <= a AND b;    -- legal (logical operation OK)
```

Despite the constraint mentioned above, there is a simple way of allowing data of type STD_LOGIC_VECTOR to participate directly in arithmetic operations. For that, the *ieee* library provides two packages, *std_logic_signed* and *std_logic_unsigned*, which allow operations with STD_LOGIC_VECTOR data to be performed as if the data were of type SIGNED or UNSIGNED, respectively.

Example: Arithmetic operations with std_logic_vector.

```
LIBRARY ieee;
USE ieee.std_logic_1164.all;
USE ieee.std_logic_unsigned.all;    -- extra package included
...
```

```
SIGNAL a: IN STD_LOGIC_VECTOR (7 DOWNTO 0);
SIGNAL b: IN STD_LOGIC_VECTOR (7 DOWNTO 0);
SIGNAL x: OUT STD_LOGIC_VECTOR (7 DOWNTO 0);
...
v <= a + b;      -- legal (arithmetic operation OK), unsigned
w <= a AND b;    -- legal (logical operation OK)
```

3.8 Data Conversion

VHDL does not allow direct operations (arithmetic, logical, etc.) between data of different types. Therefore, it is often necessary to convert data from one type to another. This can be done in basically two ways: or we write a piece of VHDL code for that, or we invoke a FUNCTION from a pre-defined PACKAGE which is capable of doing it for us.

If the data are *closely* related (that is, both operands have the same *base* type, despite being declared as belonging to two different type classes), then the *std_logic_1164* of the *ieee* library provides straightforward conversion functions. An example is shown below.

Example: Legal and illegal operations with subsets.

```
TYPE long IS INTEGER RANGE -100 TO 100;
TYPE short IS INTEGER RANGE -10 TO 10;
SIGNAL x : short;
SIGNAL y : long;
...
y <= 2*x + 5;         -- error, type mismatch
y <= long(2*x + 5);   -- OK, result converted into type long
```

Several data conversion functions can be found in the *std_logic_arith* package of the *ieee* library. They are:

• conv_integer(p) : Converts a parameter p of type INTEGER, UNSIGNED, SIGNED, or STD_ULOGIC to an INTEGER value. Notice that STD_LOGIC_ VECTOR is not included.

• conv_unsigned(p, b): Converts a parameter p of type INTEGER, UNSIGNED, SIGNED, or STD_ULOGIC to an UNSIGNED value with size b bits.

• conv_signed(p, b): Converts a parameter p of type INTEGER, UNSIGNED, SIGNED, or STD_ULOGIC to a SIGNED value with size b bits.

• conv_std_logic_vector(p, b): Converts a parameter p of type INTEGER, UN-
SIGNED, SIGNED, or STD_LOGIC to a STD_LOGIC_VECTOR value with size
b bits.

Example: Data conversion.

```
LIBRARY ieee;
USE ieee.std_logic_1164.all;
USE ieee.std_logic_arith.all;
...
SIGNAL a: IN UNSIGNED (7 DOWNTO 0);
SIGNAL b: IN UNSIGNED (7 DOWNTO 0);
SIGNAL y: OUT STD_LOGIC_VECTOR (7 DOWNTO 0);
...
y <= CONV_STD_LOGIC_VECTOR ((a+b), 8);
-- Legal operation: a+b is converted from UNSIGNED to an
-- 8-bit STD_LOGIC_VECTOR value, then assigned to y.
```

Another alternative was already mentioned in the previous section. It consists of
using the *std_logic_signed* or the *std_logic_unsigned* package from the *ieee* library.
Such packages allow operations with STD_LOGIC_VECTOR data to be performed
as if the data were of type SIGNED or UNSIGNED, respectively.

Besides the data conversion functions described above, several others are often
offered by synthesis tool vendors.

3.9   Summary

The fundamental synthesizable VHDL data types are summarized in table 3.2.

3.10   Additional Examples

We close this chapter with the presentation of additional examples illustrating the
specification and use of data types. The development of actual designs from scratch
will only be possible after we conclude laying out the basic foundations of VHDL
(chapters 1 to 4).

Example 3.1:   Dealing with Data Types

The legal and illegal assignments presented next are based on the following type
definitions and signal declarations:

Table 3.2
Synthesizable data types.

Data types	Synthesizable values
BIT, BIT_VECTOR	'0', '1'
STD_LOGIC, STD_LOGIC_VECTOR	'X', '0', '1', 'Z' (resolved)
STD_ULOGIC, STD_ULOGIC_VECTOR	'X', '0', '1', 'Z' (unresolved)
BOOLEAN	True, False
NATURAL	From 0 to +2, 147, 483, 647
INTEGER	From −2,147,483,647 to +2,147,483,647
SIGNED	From −2,147,483,647 to +2,147,483,647
UNSIGNED	From 0 to +2,147,483,647
User-defined integer type	Subset of INTEGER
User-defined enumerated type	Collection enumerated by user
SUBTYPE	Subset of any type (pre- or user-defined)
ARRAY	Single-type collection of any type above
RECORD	Multiple-type collection of any types above

```
TYPE byte IS ARRAY (7 DOWNTO 0) OF STD_LOGIC;            -- 1D
                                                        -- array
TYPE mem1 IS ARRAY (0 TO 3, 7 DOWNTO 0) OF STD_LOGIC;   -- 2D
                                                        -- array
TYPE mem2 IS ARRAY (0 TO 3) OF byte;                    -- 1Dx1D
                                                        -- array
TYPE mem3 IS ARRAY (0 TO 3) OF STD_LOGIC_VECTOR(0 TO 7); -- 1Dx1D
                                                        -- array
SIGNAL a: STD_LOGIC;                  -- scalar signal
SIGNAL b: BIT;                        -- scalar signal
SIGNAL x: byte;                       -- 1D signal
SIGNAL y: STD_LOGIC_VECTOR (7 DOWNTO 0);   -- 1D signal
SIGNAL v: BIT_VECTOR (3 DOWNTO 0);    -- 1D signal
SIGNAL z: STD_LOGIC_VECTOR (x'HIGH DOWNTO 0);  -- 1D signal
SIGNAL w1: mem1;                      -- 2D signal
SIGNAL w2: mem2;                      -- 1Dx1D signal
SIGNAL w3: mem3;                      -- 1Dx1D signal
-------- Legal scalar assignments: --------------------
x(2) <= a;          -- same types (STD_LOGIC), correct indexing
y(0) <= x(0);       -- same types (STD_LOGIC), correct indexing
z(7) <= x(5);       -- same types (STD_LOGIC), correct indexing
b <= v(3);          -- same types (BIT), correct indexing
w1(0,0) <= x(3);    -- same types (STD_LOGIC), correct indexing
```

```
w1(2,5) <= y(7);        -- same types (STD_LOGIC), correct indexing
w2(0)(0) <= x(2);       -- same types (STD_LOGIC), correct indexing
w2(2)(5) <= y(7);       -- same types (STD_LOGIC), correct indexing
w1(2,5) <= w2(3)(7);    -- same types (STD_LOGIC), correct indexing
------- Illegal scalar assignments: --------------------
b <= a;                 -- type mismatch (BIT x STD_LOGIC)
w1(0)(2) <= x(2);       -- index of w1 must be 2D
w2(2,0) <= a;           -- index of w2 must be 1Dx1D
------- Legal vector assignments: --------------------
x <= "11111110";
y <= ('1','1','1','1','1','1','0','Z');
z <= "11111" & "000";
x <= (OTHERS => '1');
y <= (7 =>'0', 1 =>'0', OTHERS => '1');
z <= y;
y(2 DOWNTO 0) <= z(6 DOWNTO 4);
w2(0)(7 DOWNTO 0) <= "11110000";
w3(2) <= y;
z <= w3(1);
z(5 DOWNTO 0) <= w3(1)(2 TO 7);
w3(1) <= "00000000";
w3(1) <= (OTHERS => '0');
w2 <= ((OTHERS=>'0'),(OTHERS=>'0'),(OTHERS=>'0'),(OTHERS=>'0'));
w3 <= ("11111100", ('0','0','0','0','Z','Z','Z','Z',),
       (OTHERS=>'0'), (OTHERS=>'0'));
w1 <= ((OTHERS=>'Z'), "11110000" ,"11110000", (OTHERS=>'0'));
------ Illegal array assignments: --------------------
x <= y;                           -- type mismatch
y(5 TO 7) <= z(6 DOWNTO 0);       -- wrong direction of y
w1 <= (OTHERS => '1');            -- w1 is a 2D array
w1(0, 7 DOWNTO 0) <="11111111";   -- w1 is a 2D array
w2 <= (OTHERS => 'Z');            -- w2 is a 1Dx1D array
w2(0, 7 DOWNTO 0) <= "11110000";  -- index should be 1Dx1D
-- Example of data type independent array initialization:
FOR i IN 0 TO 3 LOOP
   FOR j IN 7 DOWNTO 0 LOOP
      x(j) <= '0';
      y(j) <= '0'
```

```
    z(j) <= '0';
    w1(i,j) <= '0';
    w2(i)(j) <= '0';
    w3(i)(j) <= '0';
  END LOOP;
END LOOP;
```

Example 3.2:   Single Bit Versus Bit Vector

This example illustrates the difference between a single bit assignment and a bit vector assignment (that is, BIT versus BIT_VECTOR, STD_LOGIC versus STD_LOGIC_VECTOR, or STD_ULOGIC versus STD_ULOGIC_VECTOR).

Two VHDL codes are presented below. Both perform the AND operation between the input signals and assign the result to the output signal. The only difference between them is the number of bits in the input and output ports (one bit in the first, four bits in the second). The circuits inferred from these codes are shown in figure 3.2.

```
---------------------------       ------------------------------------
ENTITY and2 IS                     ENTITY and2 IS
   PORT (a, b: IN BIT;                PORT (a, b: IN BIT_VECTOR (0 TO 3);
         x: OUT BIT);                       x: OUT BIT_VECTOR (0 TO 3));
END and2;                          END and2;
---------------------------       ------------------------------------
ARCHITECTURE and2 OF and2 IS       ARCHITECTURE and2 OF and2 IS
BEGIN                              BEGIN
   x <= a AND b;                      x <= a AND b;
END and2;                          END and2;
---------------------------       ------------------------------------
```

Example 3.3:   Adder

Figure 3.3 shows the top-level diagram of a 4-bit adder. The circuit has two inputs (a, b) and one output (sum). Two solutions are presented. In the first, all signals are of type SIGNED, while in the second the output is of type INTEGER. Notice in solution 2 that a conversion function was used in line 13, for the type of $a + b$ does not match that of sum. Notice also the inclusion of the *std_logic_arith* package (line 4 of each solution), which specifies the SIGNED data type. Recall that a SIGNED value is represented like a vector; that is, similar to STD_LOGIC_VECTOR, not like an INTEGER.

```
1   ----- Solution 1: in/out=SIGNED ----------
2   LIBRARY ieee;
3   USE ieee.std_logic_1164.all;
4   USE ieee.std_logic_arith.all;
5   -------------------------------------------
6   ENTITY adder1 IS
7     PORT ( a, b : IN SIGNED (3 DOWNTO 0);
8              sum : OUT SIGNED (4 DOWNTO 0));
9   END adder1;
10  -------------------------------------------
11  ARCHITECTURE adder1 OF adder1 IS
12  BEGIN
13    sum <= a + b;
14  END adder1;
15  -------------------------------------------
```

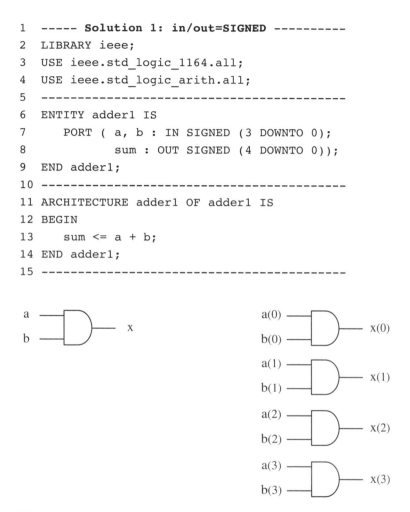

Figure 3.2
Circuits inferred from the codes of example 3.2.

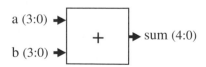

Figure 3.3
4-bit adder of example 3.3.

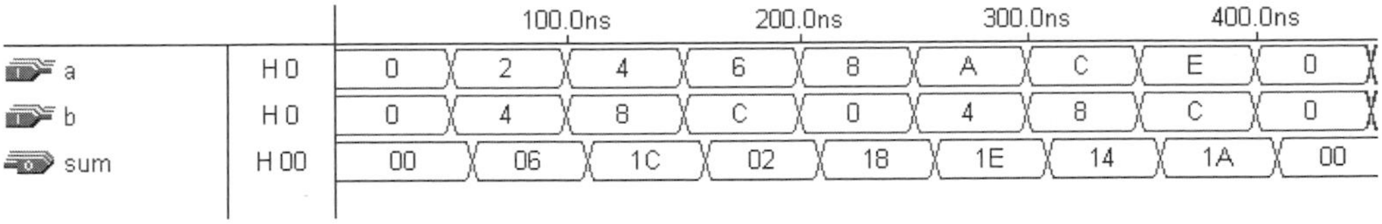

			100.0ns		200.0ns		300.0ns		400.0ns	
a	H 0	0	2	4	6	8	A	C	E	0
b	H 0	0	4	8	C	0	4	8	C	0
sum	H 00	00	06	1C	02	18	1E	14	1A	00

Figure 3.4
Simulation results of example 3.3.

```
1   ------ Solution 2: out=INTEGER -----------
2   LIBRARY ieee;
3   USE ieee.std_logic_1164.all;
4   USE ieee.std_logic_arith.all;
5   ------------------------------------------
6   ENTITY adder2 IS
7      PORT ( a, b : IN SIGNED (3 DOWNTO 0);
8              sum : OUT INTEGER RANGE -16 TO 15);
9   END adder2;
10  ------------------------------------------
11  ARCHITECTURE adder2 OF adder2 IS
12  BEGIN
13     sum <= CONV_INTEGER(a + b);
14  END adder2;
15  ------------------------------------------
```

Simulation results (for either solution) are presented in figure 3.4. Notice that the numbers are represented in hexadecimal 2's complement form. Since the input range is from -8 to 7, its representation is $7 \rightarrow 7$, $6 \rightarrow 6$, ..., $0 \rightarrow 0$, $-1 \rightarrow 15$, $-2 \rightarrow 14$, ..., $-8 \rightarrow 8$. Likewise, the output range is from -16 to 15, so its representation is $15 \rightarrow 15$, ..., $0 \rightarrow 0$, $-1 \rightarrow 31$, ..., $-16 \rightarrow 16$. Therefore, $2H + 4H = 06H$ (that is, $2 + 4 = 6$), $4H + 8H = 1CH$ (that is, $4 + (-8) = -4$), etc., where H = Hexadecimal.

3.11 Problems

The problems below are based on the following TYPE definitions and SIGNAL declarations:

```
TYPE array1 IS ARRAY (7 DOWNTO 0) OF STD_LOGIC;
TYPE array2 IS ARRAY (3 DOWNTO 0, 7 DOWNTO 0) OF STD_LOGIC;
TYPE array3 IS ARRAY (3 DOWNTO 0) OF array1;
```

```
SIGNAL a : BIT;
SIGNAL b : STD_LOGIC;:
SIGNAL x : array1;
SIGNAL y : array2;
SIGNAL w : array3;
SIGNAL z : STD_LOGIC_VECTOR (7 DOWNTO 0);
```

Problem 3.1

Determine the dimensionality (scalar, 1D, 2D, or 1Dx1D) of the signals given. Also, write down a numeric example for each signal.

Problem 3.2

Determine which among the assignments in table P3.2 are legal and which are illegal. Briefly justify your answers. Also, determine the dimensionality of each assignment (on both sides).

Problem 3.3:   Subtypes

Consider the pre-defined data types INTEGER and STD_LOGIC_VECTOR. Consider also the user-defined types ARRAY1 and ARRAY2 specified above. For each, write down a possible SUBTYPE.

Problem 3.4:   ROM

Consider the implementation of a ROM (read-only memory). It can be done utilizing a 1Dx1D CONSTANT. Say that the ROM must be organized as a pile of eight words of four bits each. Create an array called *rom*, then define a signal of type *rom* capable of solving this problem. Choose the values to be stored in the ROM and declare them along with your CONSTANT, that is, "CONSTANT my_rom: rom :=(values);".

Problem 3.5:   Simple Adder

Rewrite solution 1 of example 3.3, but this time with all input and output signals of type STD_LOGIC_VECTOR. (Suggestion: review section 3.8).

Table P3.2

Assignment	Dimension (on each side)	Legal or illegal (why)
a <= x(2);		
b <= x(2);		
b <= y(3,5);		
b <= w(5)(3);		
y(1)(0) <= z(7);		
x(0) <= y(0,0);		
x <= "1110000";		
a <= "0000000";		
y(1) <= x;		
w(0) <= y;		
w(1) <= (7=>'1', OTHERS=>'0');		
y(1) <= (0=>'0', OTHERS=>'1');		
w(2)(7 DOWNTO 0) <= x;		
w(0)(7 DOWNTO 6) <= z(5 DOWNTO 4);		
x(3) <= x(5 DOWNTO 5);		
b <= x(5 DOWNTO 5);		
y <= ((OTHERS=>'0'), (OTHERS=>'0'), (OTHERS=>'0'), "10000001");		
z(6) <= x(5);		
z(6 DOWNTO 4) <= x(5 DOWNTO 3);		
z(6 DOWNTO 4) <= y(5 DOWNTO 3);		
y(6 DOWNTO 4) <= z(3 TO 5);		
y(0, 7 DOWNTO 0) <= z;		
w(2,2) <= '1';		

4 Operators and Attributes

The purpose of this chapter, along with the preceding chapters, is to lay the basic foundations of VHDL, so in the next chapter we can start dealing with actual circuit designs. It is indeed impossible—or little productive, at least—to write any code efficiently without undertaking first the sacrifice of understanding *data types*, *operators*, and *attributes* well.

Operators and attributes constitute a relatively long list of general VHDL constructs, which are often examined only sparsely. We have collected them together in a specific chapter in order to provide a complete and more consistent view.

At the end of the chapter, a few design examples will be presented. However, due to the fact that this is still a "foundation" chapter, the examples are merely illustrative, like those in the preceding chapters. As mentioned above, we will start dealing with actual designs in chapter 5.

4.1 Operators

VHDL provides several kinds of pre-defined operators:

· Assignment operators

· Logical operators

· Arithmetic operators

· Relational operators

· Shift operators

· Concatenation operators

Each of these categories is described below.

Assignment Operators

Are used to assign values to signals, variables, and constants. They are:

<= Used to assign a value to a SIGNAL.

:= Used to assign a value to a VARIABLE, CONSTANT, or GENERIC. Used also for establishing initial values.

=> Used to assign values to individual vector elements or with OTHERS.

Example: Consider the following signal and variable declarations:

```
SIGNAL x : STD_LOGIC;
VARIABLE y : STD_LOGIC_VECTOR(3 DOWNTO 0);  -- Leftmost bit is MSB
```

```
SIGNAL w: STD_LOGIC_VECTOR( 0 TO 7);        -- Rightmost bit is
                                            -- MSB
```

Then the following assignments are legal:

```
x <= '1';        -- '1' is assigned to SIGNAL x using "<="
y := "0000";     -- "0000" is assigned to VARIABLE y using ":="
w <= "10000000";                 -- LSB is '1', the others are '0'
w <= (0 =>'1', OTHERS =>'0');    -- LSB is '1', the others are '0'
```

Logical Operators

Used to perform logical operations. The data must be of type BIT, STD_LOGIC, or STD_ULOGIC (or, obviously, their respective extensions, BIT_VECTOR, STD_LOGIC_VECTOR, or STD_ULOGIC_VECTOR). The logical operators are:

· NOT

· AND

· OR

· NAND

· NOR

· XOR

· XNOR

Notes: The NOT operator has precedence over the others. The XNOR operator was introduced in VHDL93.

Examples:

```
y <= NOT a AND b;      -- (a'.b)
y <= NOT (a AND b);    -- (a.b)'
y <= a NAND b;         -- (a.b)'
```

Arithmetic Operators

Used to perform arithmetic operations. The data can be of type INTEGER, SIGNED, UNSIGNED, or REAL (recall that the last cannot be synthesized directly). Also, if the *std_logic_signed* or the *std_logic_unsigned* package of the *ieee* library is used, then STD_LOGIC_VECTOR can also be employed directly in addition and subtraction operations (as seen in section 3.7).

+	Addition
−	Subtraction
*	Multiplication
/	Division
**	Exponentiation
MOD	Modulus
REM	Remainder
ABS	Absolute value

There are no synthesis restrictions regarding addition and subtraction, and the same is generally true for multiplication. For division, only power of two dividers (shift operation) are allowed. For exponentiation, only static values of base and exponent are accepted. Regarding the mod and rem operators, y mod x returns the remainder of y/x with the sign of x, while y rem x returns the remainder of y/x with the sign of y. Finally, abs returns the absolute value. With respect to the last three operators (mod, rem, abs), there generally is little or no synthesis support.

Comparison Operators

Used for making comparisons. The data can be of any of the types listed above. The relational (comparison) operators are:

=	Equal to
/=	Not equal to
<	Less than
>	Greater than
<=	Less than or equal to
>=	Greater than or equal to

Shift Operators

Used for shifting data. They were introduced in VHDL93. Their syntax is the following: ⟨left operand⟩ ⟨shift operation⟩ ⟨right operand⟩. The left operand must be of type BIT_VECTOR, while the right operand must be an INTEGER (+ or − in front of it is accepted). The shift operators are:

· sll   Shift left logic   – positions on the right are filled with '0's
· srl   Shift right logic   – positions on the left are filled with '0's

- sla Shift left arithmetic   – rightmost bit is replicated on the right
- sra Shift right arithmetic   – leftmost bit is replicated on the left
- rol Rotate left logic
- ror Rotate right logic

Examples: Say that x<="01001". Then:

```
y <= x sll 2;    -- logic shift to the left by 2: y<="00100"
y <= x sla 2;    -- arithmetic shift to the left by 2: y<="00111"
y <= x srl 3;    -- logic shift to the right by 3: y<="00001"
y <= x sra 3;    -- arithmetic shift to the right by 3: y<="00001"
y <= x rol 2;    -- logic rotation to the left by 2: y<="00101"
y <= x srl -2;   -- same as sll 2
```

Concatenation Operators

Are used to group values. The data can be of any of the types listed for logical operations. The concatenation operators are:

- &
- (, ,)

Examples:

```
z <= x & "1000000";    -- If x<='1', then z<="11000000"
z <= ('1','1','0','0','0','0','0','0');    -- z<="11000000"
```

4.2 Attributes

The purpose of attributes is to give VHDL more flexibility, and also allow the construction of *generic* pieces of code (code that will work for any vector or array size, for example). Such attributes are divided into two groups:

- Data attributes: Return information (a value) regarding a data vector.;
- Signal attributes: Serve to monitor a signal (return TRUE or FALSE).

In either case, to use an attribute the "'" (apostrophe) construct must be employed. Besides a list of pre-defined attributes, VHDL also allows user-defined attributes. The former will be presented in this section, while the latter will be discussed in section 4.3.

Data Attributes

The pre-defined, synthesizable data attributes are the following:

- d'LOW: Returns lower array index
- d'HIGH: Returns upper array index
- d'LEFT: Returns leftmost array index
- d'RIGHT: Returns rightmost array index
- d'LENGTH: Returns array size
- d'RANGE: Returns array range
- d'REVERSE_RANGE: Returns array range in reverse order

Example: Consider the following signal:

```
SIGNAL d : STD_LOGIC_VECTOR (7 DOWNTO 0);
```

Then:

```
d'LOW=0, d'HIGH=7, d'LEFT=7, d'RIGHT=0, d'LENGTH=8,
d'RANGE=(7 downto 0), d'REVERSE_RANGE=(0 to 7).
```

Example: Consider the following signal:

```
SIGNAL x: STD_LOGIC_VECTOR (0 TO 7);
```

Then all four LOOP statements below are synthesizable and equivalent.

```
FOR i IN 0 TO 7 LOOP ...
FOR i IN x'RANGE LOOP ...
FOR i IN x'LOW TO x'HIGH LOOP ...
FOR i IN 0 TO x'LENGTH-1 LOOP ...
```

If the signal is of *enumerated* type, then:

- d'VAL(position): Returns value in the position specified
- d'POS(value): Returns position of the value specified
- d'LEFTOF(value): Returns value in the position to the left of the value specified
- d'VAL(row, column): Returns value in the position specified; etc.

In general, there is little synthesis support for enumerated data type attributes.

Signal Attributes

In the descriptions below, s is a signal.

• s'EVENT: Returns TRUE when an event occurs on s (that is, when the value of s changes)

• s'STABLE [t]: Returns TRUE if no event has occurred on s during the optional time interval t

• s'ACTIVE: Returns TRUE when a transaction (assignment) occurs on s (value might not change)

• s'QUIET [t]: Returns TRUE if no transaction or event occurred on s during the optional time t

• s'LAST_VALUE: Returns the value of s before the last event

• s'LAST_EVENT: Returns the time elapsed since the last event of s

• s'LAST_ACTIVE: Returns the time elapsed since the last transaction (assignment) of s.

Though the attributes above are used mainly for simulations, the first two are synthesizable, s'EVENT being the most often used of them all (useful to infer registers).

Example: All four assignments shown below are synthesizable and equivalent. They return TRUE when an event (a change) occurs on clk, *and* the final value of clk is '1' (that is, when a rising edge occurs on clk).

```
IF (clk'EVENT AND clk='1')...        -- EVENT attribute used
                                        with IF
IF (NOT clk'STABLE AND clk='1')...   -- STABLE attribute used
                                        with IF
WAIT UNTIL (clk'EVENT AND clk='1');  -- EVENT attribute used
                                        with WAIT
IF RISING_EDGE(clk)...               -- a function call
```

4.3   User-Defined Attributes

We saw above attributes of the type HIGH, RANGE, EVENT, etc. Those are all pre-defined in VHDL87. However, VHDL also allows the construction of user-defined attributes.

To employ a user-defined attribute, it must be *declared* and *specified*. The syntax is the following:

Attribute declaration:

```
ATTRIBUTE attribute_name: attribute_type;
```

Attribute specification:

```
ATTRIBUTE attribute_name OF target_name: class IS value;
```

where:

attribute_type: any data type (BIT, INTEGER, STD_LOGIC_VECTOR, etc.)

class: TYPE, SIGNAL, FUNCTION, etc.

value: '0', 27, "00 11 10 01", etc.

Example:

```
ATTRIBUTE number_of_inputs: INTEGER;              -- declaration
ATTRIBUTE number_of_inputs OF nand3: SIGNAL IS 3; -- specification
    ...
inputs <= nand3'number_of_pins;    -- attribute call, returns 3
```

Example: Enumerated encoding.

A popular user-defined attribute, which is provided by synthesis tool vendors, is the *enum_encoding* attribute. By default, enumerated data types are encoded sequentially. Thus, if we consider the enumerated data type `color` shown below:

```
TYPE color IS (red, green, blue, white);
```

its states will be encoded as red = "00", green = "01", blue = "10", and white = "11". Enum_encoding allows the default encoding (sequential) to be changed. Thus the following encoding scheme could be employed, for example:

```
ATTRIBUTE enum_encoding OF color: TYPE IS "11 00 10 01";
```

A user-defined attribute can be declared anywhere, except in a PACKAGE
BODY. When not recognized by the synthesis tool, it is simply ignored, or a warning
is issued.

4.4   Operator Overloading

We have just seen that attributes can be user-defined. The same is true for operators.
As an example, let us consider the pre-defined arithmetic operators seen in section
4.1 (+, −, *, /, etc.). They specify arithmetic operations between data of certain types
(INTEGER, for example). For instance, the pre-defined "+" operator does not
allow addition between data of type BIT.

We can define our own operators, using the same *name* as the pre-defined ones.
For example, we could use "+" to indicate a new kind of addition, this time between
values of type BIT_VECTOR. This technique is called *operator overloading*.

Example: Consider that we want to add an integer to a binary 1-bit number. Then
the following FUNCTION could be used (details on how to construct and use a
FUNCTION will be seen in chapter 11):

```
---------------------------------------
FUNCTION "+" (a: INTEGER, b: BIT) RETURN INTEGER IS
BEGIN
    IF (b='1') THEN RETURN a+1;
    ELSE RETURN a;
    END IF;
END "+";
---------------------------------------
```

A call to the function above could thus be the following:

```
-----------------------------
SIGNAL inp1, outp: INTEGER RANGE 0 TO 15;
SIGNAL inp2: BIT;
    (...)
outp <= 3 + inp1 + inp2;
    (...)
-----------------------------
```

In "outp<=3+inp1+inp2;", the first "+" is the pre-defined addition operator
(adds two integers), while the second is the overloaded user-defined addition operator
(adds an integer and a bit).

4.5   GENERIC

As the name suggests, GENERIC is a way of specifying a *generic* parameter (that is, a static parameter that can be easily modified and adapted to different applications). The purpose is to confer the code more flexibility and reusability.

A GENERIC statement, when employed, must be declared in the ENTITY. The specified parameter will then be truly global (that is, visible to the whole design, including the ENTITY itself). Its syntax is shown below.

```
GENERIC (parameter_name : parameter_type := parameter_value);
```

Example: The GENERIC statement below specifies a parameter called n, of type INTEGER, whose default value is 8. Therefore, whenever n is found in the ENTITY itself or in the ARCHITECTURE (one or more) that follows, its value will be assumed to be 8.

```
ENTITY my_entity IS
   GENERIC (n : INTEGER := 8);
   PORT (...);
END my_entity;
```

More than one GENERIC parameter can be specified in an ENTITY. For example:

```
GENERIC (n: INTEGER := 8;
         vector: BIT_VECTOR (7 DOWNTO 0) := "00001111");
```

Complete design examples, further illustrating the use of GENERIC and other attributes and operators, are presented below.

4.6   Examples

We show now a few complete design examples, with the purpose of further illustrating the use of operators, attributes and GENERIC. Recall, however, that so far we have just worked on establishing the basic foundations of VHDL, with the formal discussion on coding techniques starting only in the next chapter (chapter 5). Therefore, a first-time VHDL student should not feel discouraged if the constructs in the examples look still unfamiliar. Instead, you may have a look at

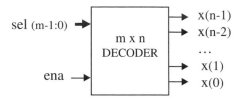

ena	sel	x
0	XX	1111
1	00	1110
	01	1101
	10	1011
	11	0111

Figure 4.1
Decoder of example 4.1.

the examples now, and then, after studying chapters 5 to 7, return and reexamine them.

Example 4.1:   Generic Decoder

Figure 4.1 shows the top-level diagram of a generic m-by-n decoder. The circuit has two inputs, sel (m bits) and ena (single bit), and one output, x (n bits). We assume that n is a power of two, so $m = \log_2 n$. If ena = '0', then all bits of x should be high; otherwise, the output bit selected by sel should be low, as illustrated in the truth table of figure 4.1.

The ARCHITECTURE below is totally generic, for the only changes needed to operate with different values of m and n are in the ENTITY (through sel, line 7, and x, line 8, respectively). In this example, we have used m = 3 and n = 8. However, though this works fine, the use of GENERIC would have made it clearer that m and n are indeed *generic* parameters. That is indeed the procedure that we will adopt in the other examples that follow (please refer to problem 4.4).

Notice in the code below the use of the following *operators*: "+" (line 22), "*" (lines 22 and 24), ":=" (lines 17, 18, 22, 24, and 27), "<=" (line 29), and "=>" (line 17). Notice also the use of the following *attributes*: HIGH (lines 14–15) and RANGE (line 20).

```
1   ---------------------------------------------
2   LIBRARY ieee;
3   USE ieee.std_logic_1164.all;
4   ---------------------------------------------
5   ENTITY decoder IS
6      PORT ( ena : IN STD_LOGIC;
7                 sel : IN STD_LOGIC_VECTOR (2 DOWNTO 0);
8                 x : OUT STD_LOGIC_VECTOR (7 DOWNTO 0));
9   END decoder;
```

Figure 4.2
Simulation results of example 4.1.

```
10 ----------------------------------------------
11 ARCHITECTURE generic_decoder OF decoder IS
12 BEGIN
13    PROCESS (ena, sel)
14       VARIABLE temp1 : STD_LOGIC_VECTOR (x'HIGH DOWNTO 0);
15       VARIABLE temp2 : INTEGER RANGE 0 TO x'HIGH;
16    BEGIN
17       temp1 := (OTHERS => '1');
18       temp2 := 0;
19       IF (ena='1') THEN
20          FOR i IN sel'RANGE LOOP   -- sel range is 2 downto 0
21             IF (sel(i)='1') THEN   -- Bin-to-Integer conversion
22                temp2:=2*temp2+1;
23             ELSE
24                temp2 := 2*temp2;
25             END IF;
26          END LOOP;
27          temp1(temp2):='0';
28       END IF;
29       x <= temp1;
30    END PROCESS;
31 END generic_decoder;
32 ----------------------------------------------
```

The functionality of the encoder above can be verified in the simulation results of figure 4.2. As can be seen, all outputs are high, that is, x = "11111111" (decimal 255), when ena = '0'. After ena has been asserted, only one output bit (that selected by sel) is turned low. For example, when sel = "000" (decimal 0), x = "11111110" (decimal 254); when sel = "001" (decimal 1), x = "11111101" (decimal 253); when sel = "010" (decimal 2), x = "11111011" (decimal 251); and so on.

Figure 4.3
Generic parity detector of example 4.2.

| | | 125.0ns | 250.0ns | 375.0ns | 500.0ns | 625.0ns | 7 |

input D 0 | 0 | 1 | 2 | 3 | 4 | 5 | 6 | 7

output 0

Figure 4.4
Simulation results of example 4.2.

Example 4.2:   Generic Parity Detector

Figure 4.3 shows the top-level diagram of a parity detector. The circuit must provide output = '0' when the number of '1's in the input vector is even, or output = '1' otherwise. Notice in the VHDL code below that the ENTITY contains a GENERIC statement (line 3), which defines n as 7. This code would work for any other vector size, being only necessary to change the value of n in that line. You are invited to highlight the *operators* and *attributes* that appear in this design.

```
1    ---------------------------------------------
2    ENTITY parity_det IS
3       GENERIC (n : INTEGER := 7);
4       PORT ( input: IN BIT_VECTOR (n DOWNTO 0);
5               output: OUT BIT);
6    END parity_det;
7    ---------------------------------------------
8    ARCHITECTURE parity OF parity_det IS
9    BEGIN
10      PROCESS (input)
11         VARIABLE temp: BIT;
12      BEGIN
13         temp := '0';
14         FOR i IN input'RANGE LOOP
```

Figure 4.5
Generic parity generator of example 4.3.

```
15              temp := temp XOR input(i);
16          END LOOP;
17          output <= temp;
18      END PROCESS;
19  END parity;
20  --------------------------------------------
```

Simulation results from the circuit synthesized with the code above are shown in figure 4.4. Notice that when input = "00000000" (decimal 0), the output is '0', because the number of '1's is even; when input = "00000001" (decimal 1), the output is '1', because the number of '1's is odd; and so on.

Example 4.3:   Generic Parity Generator

The circuit of figure 4.5 must add one bit to the input vector (on its left). Such bit must be a '0' if the number of '1's in the input vector is even, or a '1' if it is odd, such that the resulting vector will always contain an even number of '1's (even parity).

A VHDL code for the parity generator is shown below. Once again, you are invited to highlight the operators and attributes used in the design.

```
1   --------------------------------------------
2   ENTITY parity_gen IS
3       GENERIC (n : INTEGER := 7);
4       PORT ( input: IN BIT_VECTOR (n-1 DOWNTO 0);
5              output: OUT BIT_VECTOR (n DOWNTO 0));
6   END parity_gen;
7   --------------------------------------------
8   ARCHITECTURE parity OF parity_gen IS
9   BEGIN
10      PROCESS (input)
11          VARIABLE temp1: BIT;
```

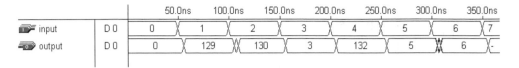

Figure 4.6
Simulation results of example 4.3.

Table 4.1
Operators.

Operator type	Operators	Data types
Assignment	<=, :=, =>	Any
Logical	NOT, AND, NAND, OR, NOR, XOR, XNOR	BIT, BIT_VECTOR, STD_LOGIC, STD_LOGIC_VECTOR, STD_ULOGIC, STD_ULOGIC_VECTOR
Arithmetic	+, −, *, /, ** (mod, rem, abs)♦	INTEGER, SIGNED, UNSIGNED
Comparison	=, /=, <, >, <=, >=	All above
Shift	sll, srl, sla, sra, rol, ror	BIT_VECTOR
Concatenation	&, (, , ,)	Same as for logical operators, plus SIGNED and UNSIGNED

```
12          VARIABLE temp2: BIT_VECTOR (output'RANGE);
13      BEGIN
14          temp1 := '0';
15          FOR i IN input'RANGE LOOP
16              temp1 := temp1 XOR input(i);
17              temp2(i) := input(i);
18          END LOOP;
19          temp2(output'HIGH) := temp1;
20          output <= temp2;
21      END PROCESS;
22  END parity;
23  ------------------------------------------------
```

Simulation results are presented in figure 4.6. As can be seen, when input = "0000000" (decimal 0, with seven bits), output = "00000000" (decimal 0, with eight bits); when input = "0000001" (decimal 1, with seven bits), output = "10000001" (decimal 129, with eight bits); and so on.

Table 4.2
Attributes.

Application	Attributes	Return value
For regular DATA	d'LOW	Lower array index
	d'HIGH	Upper array index
	d'LEFT	Leftmost array index
	d'RIGHT	Rightmost array index
	d'LENGTH	Vector size
	d'RANGE	Vector range
	d'REVERSE_RANGE	Reverse vector range
For enumerated	d'VAL(pos)♦	Value in the position specified
DATA	d'POS(value)♦	Position of the value specified
	d'LEFTOF(value)♦	Value in the position to the left of the value specified
	d'VAL(row, column)♦	Value in the position specified
For a SIGNAL	s'EVENT	True when an event occurs on s
	s'STABLE	True if no event has occurred on s
	s'ACTIVE♦	True if s is high

4.7   Summary

A summary of VHDL operators and attributes is presented in tables 4.1 and 4.2, respectively. The constructs that are not synthesizable (or have little synthesis support) are marked with the "♦" symbol.

4.8   Problems

Problems 4.1 to 4.3 are based on the following signal declarations:

```
SIGNAL a : BIT := '1';
SIGNAL b : BIT_VECTOR (3 DOWNTO 0) := "1100";
SIGNAL c : BIT_VECTOR (3 DOWNTO 0) := "0010";
SIGNAL d : BIT_VECTOR (7 DOWNTO 0);
SIGNAL e : INTEGER RANGE 0 TO 255;
SIGNAL f : INTEGER RANGE -128 TO 127;
```

Problem 4.1:   Operators (fill in the blanks)

```
x1 <= a & c;          ->     x1 <= _____
x2 <= c & b;          ->     x2 <= _____
x3 <= b XOR c;        ->     x3 <= _____
x4 <= a NOR b(3);     ->     x4 <= _____
```

```
x5 <= b sll 2;          ->      x5 <= _____
x6 <= b sla 2;          ->      x6 <= _____
x7 <= b rol 2;          ->      x7 <= _____
x8 <= a AND NOT b(0) AND NOT c(1);    ->    x8 <= _____
d <= (5=>'0', OTHERS=>'1');           ->    d <= _____
```

Problem 4.2:   Attributes (fill in the blanks)

```
c'LOW                   ->      _____
d'HIGH                  ->      _____
c'LEFT                  ->      _____
d'RIGHT                 ->      _____
c'RANGE                 ->      _____
d'LENGTH                ->      _____
c'REVERSE_RANGE         ->      _____
```

Problem 4.3:   Legal and Illegal Operations

Verify whether each of the operations below is legal or illegal. Briefly justify your answers.

```
b(0) AND a
a + d(7)
NOT b XNOR c
c + d
e - f
IF (b<c) ...
IF (b>=a) ...
IF (f/=e) ...
IF (e>d) ...
b sra 1
c srl -2
f ror 3
e*3
5**5
f/4
e/3
d <= c
d(6 DOWNTO 3) := b
e <= d
f := 100
```

Problem 4.4:   Generic Decoder

The questions below are related to the decoder circuit designed in example 4.1.

(a) In order for that design to operate with another vector size, two values must be changed: the range of sel (line 7) and the range of x (line 8). We want now to transform that design in a truly generic one. In order to do so, introduce a GENERIC statement in the ENTITY, specifying the number of bits of sel (say, n = 3), then replace the upper range limits of sel and x by a function of n. Synthesize and simulate your circuit in order to verify its functionality.

(b) In example 4.1, a binary-to-integer conversion was implemented (lines 20–26). This conversion could be avoided if sel had been declared as an INTEGER. Modify the code, declaring sel as an INTEGER. The code should remain truly generic, so the range of sel must be specified in terms of n. Synthesize and simulate your new code.

Problem 4.5

List all operators, attributes and generics employed in examples 4.2 and 4.3.

5 Concurrent Code

Having finished laying out the basic foundations of VHDL (chapters 1 to 4), we can now concentrate on the design (code) itself.

VHDL code can be *concurrent* (parallel) or *sequential*. The former will be studied in this chapter, while the latter will be seen in chapter 6. This division is very important, for it allows a better understanding of which statements are intended for each kind of code, as well as the consequences of using one or the other.

The concurrent statements in VHDL are WHEN and GENERATE. Besides them, assignments using only operators (AND, NOT, +, *, sll, etc.) can also be used to construct concurrent code. Finally, a special kind of assignment, called BLOCK, can also be employed in this kind of code.

5.1   Concurrent versus Sequential

We start this chapter by reviewing the fundamental differences between *combinational* logic and *sequential* logic, and by contrasting them with the differences between *concurrent* code and *sequential* code.

Combinational versus Sequential *Logic*

By definition, *combinational* logic is that in which the output of the circuit depends solely on the current inputs (figure 5.1(a)). It is then clear that, in principle, the system requires no memory and can be implemented using conventional logic gates.

In contrast, *sequential* logic is defined as that in which the output does depend on previous states (figure 5.1(b)). Therefore, storage elements are required, which are connected to the combinational logic block through a *feedback* loop, such that now the stored states (created by previous inputs) will also affect the output of the circuit.

A common mistake is to think that any circuit that possesses storage elements (flip-flops) is sequential. A RAM (Random Access Memory) is an example. A RAM can be modeled as in figure 5.2. Notice that the storage elements appear in a *forward* path rather than in a feedback loop. The memory-read operation depends only on the address vector presently applied to the RAM input, with the retrieved value having nothing to do with previous memory accesses.

Concurrent versus Sequential *Code*

VHDL code is inherently *concurrent* (parallel). Only statements placed inside a PROCESS, FUNCTION, or PROCEDURE are *sequential*. Still, though within these blocks the execution is sequential, the block, as a whole, is concurrent with any other (external) statements. Concurrent code is also called *dataflow* code.

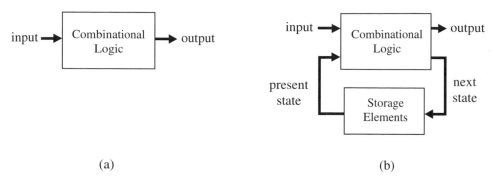

(a)   (b)

Figure 5.1
Combinational (a) versus sequential (b) logic.

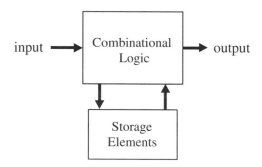

Figure 5.2
RAM model.

As an example, let us consider a code with three concurrent statements (stat1, stat2, stat3). Then any of the alternatives below will render the same physical circuit:

stat1   stat3   stat1

stat2   ≡   stat2   ≡   stat3   ≡   etc.

stat3   stat1   stat2

It is then clear that, since the order does not matter, purely concurrent code can not be used to implement synchronous circuits (the only exception is when a GUARDED BLOCK is used). In other words, in general we can only build *combinational* logic circuits with concurrent code. To obtain *sequential* logic circuits, *sequential* code (chapter 6) must be employed. Indeed, with the latter we can implement both, sequential as well as combinational circuits.

In this chapter, we will discuss *concurrent* code, that is, we will study the statements that can only be used outside PROCESSES, FUNCTIONS, or PROCEDURES. They are the WHEN statement and the GENERATE statement. Besides them, assignments using only operators (logical, arithmetic, etc) can obviously also be used to create combinational circuits. Finally, a special kind of statement, called BLOCK, can also be employed.

In summary, in concurrent code the following can be used:

• Operators;
• The WHEN statement (WHEN/ELSE or WITH/SELECT/WHEN);
• The GENERATE statement;
• The BLOCK statement.

Each of these cases is described below.

5.2   Using Operators

This is the most basic way of creating concurrent code. Operators (AND, OR, +, −. *, sll, sra, etc.) were discussed in section 4.1, being a summary repeated in table 5.1 below.

Operators can be used to implement any combinational circuit. However, as will become apparent later, complex circuits are usually easier to write using sequential code, even if the circuit does not contain sequential logic. In the example that follows, a design using only logical operators is presented.

Table 5.1
Operators.

Operator type	Operators	Data types
Logical	NOT, AND, NAND, OR, NOR, XOR, XNOR	BIT, BIT_VECTOR, STD_LOGIC, STD_LOGIC_VECTOR, STD_ULOGIC, STD_ULOGIC_VECTOR
Arithmetic	+, −, *, /, ** (mod, rem, abs)	INTEGER, SIGNED, UNSIGNED
Comparison	=, /=, <, >, <=, >=	All above
Shift	sll, srl, sla, sra, rol, ror	BIT_VECTOR
Concatenation	&, (, ,)	Same as for logical operators, plus SIGNED and UNSIGNED

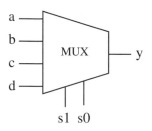

Figure 5.3
Multiplexer of example 5.1.

Example 5.1:   Multiplexer #1

Figure 5.3 shows a 4-input, one bit per input multiplexer. The output must be equal to the input selected by the selection bits, s1-s0. Its implementation, using only logical operators, can be done as follows:

```
1   ----------------------------------------
2   LIBRARY ieee;
3   USE ieee.std_logic_1164.all;
4   ----------------------------------------
5   ENTITY mux IS
6      PORT ( a, b, c, d, s0, s1: IN STD_LOGIC;
7               y: OUT STD_LOGIC);
8   END mux;
9   ----------------------------------------
10  ARCHITECTURE pure_logic OF mux IS
11  BEGIN
12     y <=  (a AND NOT s1 AND NOT s0) OR
13            (b AND NOT s1 AND s0) OR
14            (c AND s1 AND NOT s0) OR
15            (d AND s1 AND s0);
16  END pure_logic;
17  ----------------------------------------
```

Simulation results, confirming the functionality of the circuit, are shown in figure 5.4.

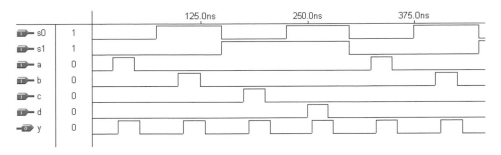

Figure 5.4
Simulation results of example 5.1.

5.3   WHEN (Simple and Selected)

As mentioned above, WHEN is one of the fundamental concurrent statements (along with operators and GENERATE). It appears in two forms: WHEN / ELSE (*simple* WHEN) and WITH / SELECT / WHEN (*selected* WHEN). Its syntax is shown below.

WHEN / ELSE:

```
assignment WHEN condition ELSE
assignment WHEN condition ELSE
...;
```

WITH / SELECT / WHEN:

```
WITH identifier SELECT
assignment WHEN value,
assignment WHEN value,
...;
```

Whenever WITH / SELECT / WHEN is used, all permutations must be tested, so the keyword OTHERS is often useful. Another important keyword is UN-AFFECTED, which should be used when no action is to take place.

Example:

```
------ With WHEN/ELSE ------------------------
outp <= "000" WHEN (inp='0' OR reset='1') ELSE
        "001" WHEN ctl='1' ELSE
        "010";

---- With WITH/SELECT/WHEN --------------------
WITH control SELECT
   output <= "000" WHEN reset,
             "111" WHEN set,
             UNAFFECTED WHEN OTHERS;
----------------------------------------------
```

Another important aspect related to the WHEN statement is that the "WHEN value" shown in the syntax above can indeed take up three forms:

```
WHEN value                 -- single value
WHEN value1 to value2      -- range, for enumerated data types
                           -- only
WHEN value1 | value2 |...  -- value1 or value2 or ...
```

Example 5.2:   Multiplexer #2

This example shows the implementation of the same multiplexer of example 5.1, but with a slightly different representation for the sel input (figure 5.5). However, in it WHEN was employed instead of logical operators. Two solutions are presented: one using WHEN/ELSE (simple WHEN) and the other with WITH/SELECT/WHEN (selected WHEN). The experimental results are obviously similar to those obtained in example 5.1.

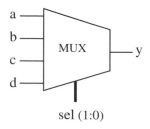

Figure 5.5
Multiplexer of example 5.2.

```
1    ------- Solution 1: with WHEN/ELSE --------
2    LIBRARY ieee;
3    USE ieee.std_logic_1164.all;
4    ---------------------------------------------
5    ENTITY mux IS
6       PORT ( a, b, c, d: IN STD_LOGIC;
7                sel: IN STD_LOGIC_VECTOR (1 DOWNTO 0);
8                y: OUT STD_LOGIC);
9    END mux;
10   ---------------------------------------------
11   ARCHITECTURE mux1 OF mux IS
12   BEGIN
13      y <=  a WHEN sel="00" ELSE
14            b WHEN sel="01" ELSE
15            c WHEN sel="10" ELSE
16            d;
17   END mux1;
18   ---------------------------------------------

1    --- Solution 2: with WITH/SELECT/WHEN -----
2    LIBRARY ieee;
3    USE ieee.std_logic_1164.all;
4    ---------------------------------------------
5    ENTITY mux IS
6       PORT ( a, b, c, d: IN STD_LOGIC;
7                sel: IN STD_LOGIC_VECTOR (1 DOWNTO 0);
8                y: OUT STD_LOGIC);
9    END mux;
10   ---------------------------------------------
11   ARCHITECTURE mux2 OF mux IS
12   BEGIN
13      WITH sel SELECT
14         y <=  a WHEN "00",      -- notice "," instead of ";"
15               b WHEN "01",
16               c WHEN "10",
17               d WHEN OTHERS;    -- cannot be "d WHEN "11" "
18   END mux2;
19   ---------------------------------------------
```

In the solutions above, sel could have been declared as an INTEGER, in which case the code would be the following:

```
1   ----------------------------------------------
2   LIBRARY ieee;
3   USE ieee.std_logic_1164.all;
4   ----------------------------------------------
5   ENTITY mux IS
6      PORT ( a, b, c, d: IN STD_LOGIC;
7              sel: IN INTEGER RANGE 0 TO 3;
8              y: OUT STD_LOGIC);
9   END mux;
10  ---- Solution 1: with WHEN/ELSE ---------------
11  ARCHITECTURE mux1 OF mux IS
12  BEGIN
13     y <=  a WHEN sel=0 ELSE
14           b WHEN sel=1 ELSE
15           c WHEN sel=2 ELSE
16           d;
17  END mux1;
18  -- Solution 2: with WITH/SELECT/WHEN  --------
19  ARCHITECTURE mux2 OF mux IS
20  BEGIN
21     WITH sel SELECT
22        y <=  a WHEN 0,
23              b WHEN 1,
24              c WHEN 2,
25              d WHEN 3;   -- here, 3 or OTHERS are equivalent,
26  END mux2;               -- for all options are tested anyway
27  ----------------------------------------------
```

Note: Only one ARCHITECTURE can be synthesized at a time. Therefore, whenever we show more than one solution within the same overall code (like above), it is implicit that all solutions but one must be commented out (with "--"), or a synthesis script must be used, in order to synthesize the remaining solution. In simulations, the CONFIGURATION statement can be used to select a specific architecture.

Note: For a *generic* mux, please refer to problem 5.1.

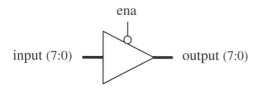

Figure 5.6
Tri-state buffer of example 5.3.

Example 5.3:   Tri-state Buffer

This is another example that illustrates the use of WHEN. The 3-state buffer of figure 5.6 must provide output = input when ena (enable) is low, or output = "ZZZZZZZZ" (high impedance) otherwise.

```
1   LIBRARY ieee;
2   USE ieee.std_logic_1164.all;
3   -----------------------------------------------
4   ENTITY tri_state IS
5      PORT ( ena: IN STD_LOGIC;
6              input: IN STD_LOGIC_VECTOR (7 DOWNTO 0);
7              output: OUT STD_LOGIC_VECTOR (7 DOWNTO 0));
8   END tri_state;
9   -----------------------------------------------
10  ARCHITECTURE tri_state OF tri_state IS
11  BEGIN
12     output <= input WHEN (ena='0') ELSE
13              (OTHERS => 'Z');
14  END tri_state;
15  -----------------------------------------------
```

Simulation results from the circuit synthesized with the code above are shown in figure 5.7. As expected, the output stays in the high-impedance state while ena is high, being a copy of the input when ena is turned low.

Example 5.4:   Encoder

The top-level diagram of an n-by-m encoder is shown in figure 5.8. We assume that n is a power of two, so m = $\log_2$n. One and only one input bit is expected to be high at a time, whose address must be encoded at the output. Two solutions are presented, one using WHEN / ELSE, and the other with WITH / SELECT / WHEN.

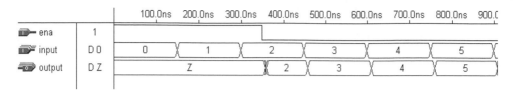

Figure 5.7
Simulation results of example 5.3.

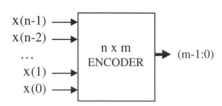

Figure 5.8
Encoder of example 5.4.

```
1   ---- Solution 1: with WHEN/ELSE -------------
2   LIBRARY ieee;
3   USE ieee.std_logic_1164.all;
4   ------------------------------------------------
5   ENTITY encoder IS
6      PORT ( x: IN STD_LOGIC_VECTOR (7 DOWNTO 0);
7             y: OUT STD_LOGIC_VECTOR (2 DOWNTO 0));
8   END encoder;
9   ------------------------------------------------
10  ARCHITECTURE encoder1 OF encoder IS
11  BEGIN
12     y <=   "000" WHEN x="00000001" ELSE
13            "001" WHEN x="00000010" ELSE
14            "010" WHEN x="00000100" ELSE
15            "011" WHEN x="00001000" ELSE
16            "100" WHEN x="00010000" ELSE
17            "101" WHEN x="00100000" ELSE
18            "110" WHEN x="01000000" ELSE
19            "111" WHEN x="10000000" ELSE
20            "ZZZ";
```

```
21 END encoder1;
22 ------------------------------------------------

1  ---- Solution 2: with WITH/SELECT/WHEN ------
2  LIBRARY ieee;
3  USE ieee.std_logic_1164.all;
4  ------------------------------------------------
5  ENTITY encoder IS
6     PORT ( x: IN STD_LOGIC_VECTOR (7 DOWNTO 0);
7             y: OUT STD_LOGIC_VECTOR (2 DOWNTO 0));
8  END encoder;
9  ------------------------------------------------
10 ARCHITECTURE encoder2 OF encoder IS
11 BEGIN
12    WITH x SELECT
13       y <=   "000" WHEN "00000001",
14              "001" WHEN "00000010",
15              "010" WHEN "00000100",
16              "011" WHEN "00001000",
17              "100" WHEN "00010000",
18              "101" WHEN "00100000",
19              "110" WHEN "01000000",
20              "111" WHEN "10000000",
21              "ZZZ" WHEN OTHERS;
22 END encoder2;
23 ------------------------------------------------
```

Notice that the code above has a long test list (lines 12–20 in solution 1, lines 13–21 in solution 2). The situation becomes even more cumbersome when the number of selection bits grows. In such a case, the GENERATE statement (section 5.4) or the LOOP statement (section 6.6) can be employed.

Simulation results (from either solution) are shown in figure 5.9.

Example 5.5:   ALU

An ALU (Arithmetic Logic Unit) is shown in figure 5.10. As the name says, it is a circuit capable of executing both kinds of operations, arithmetic as well as logical. Its operation is described in the truth table of figure 5.10. The output (arithmetic or logical) is selected by the MSB of sel, while the specific operation is selected by sel's other three bits.

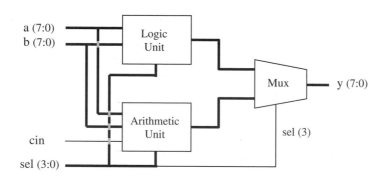

Figure 5.9
Simulation results of example 5.4.

sel	Operation	Function	Unit
0000	y <= a	Transfer a	
0001	y <= a+1	Increment a	
0010	y <= a-1	Decrement a	
0011	y <= b	Transfer b	Arithmetic
0100	y <= b+1	Increment b	
0101	y <= b-1	Decrement b	
0110	y <= a+b	Add a and b	
0111	y <= a+b+cin	Add a and b with carry	
1000	y <= NOT a	Complement a	
1001	y <= NOT b	Complement b	
1010	y <= a AND b	AND	
1011	y <= a OR b	OR	Logic
1100	y <= a NAND b	NAND	
1101	y <= a NOR b	NOR	
1110	y <= a XOR b	XOR	
1111	y <= a XNOR b	XNOR	

Figure 5.10
ALU of example 5.5.

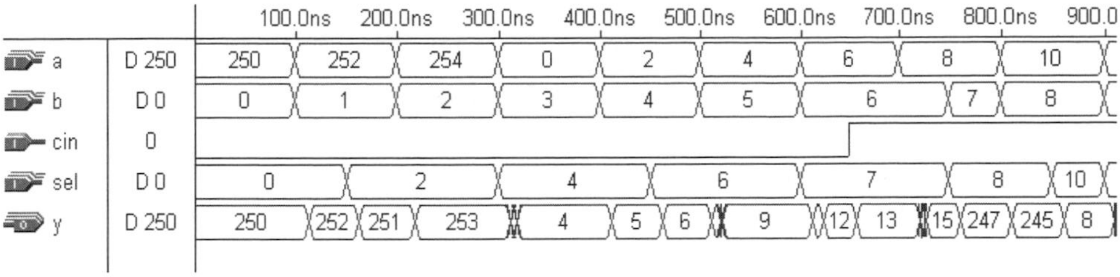

Figure 5.11
Simulation results of example 5.5.

The solution presented below, besides using only concurrent code, also illustrates the use of the same data type to perform both arithmetic and logical operations. That is possible due to the presence of the *std_logic_unsigned* package of the *ieee* library (discussed in section 3.7). Two signals, arith and logic, are used to hold the results from the arithmetic and logic units, respectively, being the value passed to the output selected by the multiplexer. Simulation results are shown in figure 5.11.

```
1   --------------------------------------------
2   LIBRARY ieee;
3   USE ieee.std_logic_1164.all;
4   USE ieee.std_logic_unsigned.all;
5   --------------------------------------------
6   ENTITY ALU IS
7      PORT (a, b: IN STD_LOGIC_VECTOR (7 DOWNTO 0);
8                  sel: IN STD_LOGIC_VECTOR (3 DOWNTO 0);
9                  cin: IN STD_LOGIC;
10                 y: OUT STD_LOGIC_VECTOR (7 DOWNTO 0));
11  END ALU;
12  --------------------------------------------
13  ARCHITECTURE dataflow OF ALU IS
14     SIGNAL arith, logic: STD_LOGIC_VECTOR (7 DOWNTO 0);
15  BEGIN
16     ----- Arithmetic unit: ------
17     WITH sel(2 DOWNTO 0) SELECT
18        arith <=  a WHEN "000",
19                  a+1 WHEN "001",
20                  a-1 WHEN "010",
21                  b WHEN "011",
22                  b+1 WHEN "100",
```

```
23                    b-1 WHEN "101",
24                    a+b WHEN "110",
25                    a+b+cin WHEN OTHERS;
26     ----- Logic unit: -----------
27     WITH sel(2 DOWNTO 0) SELECT
28        logic <=  NOT a WHEN "000",
29                  NOT b WHEN "001",
30                  a AND b WHEN "010",
31                  a OR b WHEN "011",
32                  a NAND b WHEN "100",
33                  a NOR b WHEN "101",
34                  a XOR b WHEN "110",
35                  NOT (a XOR b) WHEN OTHERS;
36     -------- Mux: ---------------
37     WITH sel(3) SELECT
38        y <=  arith WHEN '0',
39              logic WHEN OTHERS;
40  END dataflow;
41  ----------------------------------------------
```

5.4 GENERATE

GENERATE is another *concurrent* statement (along with operators and WHEN). It is equivalent to the *sequential* statement LOOP (chapter 6) in the sense that it allows a section of code to be repeated a number of times, thus creating several instances of the same assignments. Its regular form is the FOR / GENERATE construct, with the syntax shown below. Notice that GENERATE must be labeled.

FOR / GENERATE:

```
label: FOR identifier IN range GENERATE
   (concurrent assignments)
END GENERATE;
```

An irregular form is also available, which uses IF/GENERATE (with an IF equivalent; recall that originally IF is a sequential statement). Here ELSE is not allowed. In the same way that IF/GENERATE can be nested inside FOR/GENERATE (syntax below), the opposite can also be done.

IF / GENERATE nested inside FOR / GENERATE:

```
label1: FOR identifier IN range GENERATE
    ...
    label2: IF condition GENERATE
        (concurrent assignments)
    END GENERATE;
    ...
END GENERATE;
```

Example:

```
SIGNAL x: BIT_VECTOR (7 DOWNTO 0);
SIGNAL y: BIT_VECTOR (15 DOWNTO 0);
SIGNAL z: BIT_VECTOR (7 DOWNTO 0);
...
G1: FOR i IN x'RANGE GENERATE
   z(i) <= x(i) AND y(i+8);
END GENERATE;
```

One important remark about GENERATE (and the same is true for LOOP, which will be seen in chapter 6) is that both limits of the range must be static. As an example, let us consider the code below, where choice is an input (non-static) parameter. This kind of code is generally not synthesizable.

```
NotOK: FOR i IN 0 TO choice GENERATE
   (concurrent statements)
END GENERATE;
```

We also must be aware of multiply-driven (unresolved) signals. For example,

```
OK: FOR i IN 0 TO 7 GENERATE
   output(i)<='1' WHEN (a(i) AND b(i))='1' ELSE '0';
END GENERATE;
```

is fine. However, the compiler will complain that accum is multiply driven (and stop compilation) in either of the following two cases:

```
NotOK: FOR i IN 0 TO 7 GENERATE
   accum <="11111111" WHEN (a(i) AND b(i))='1' ELSE "00000000";
END GENERATE;
```

```
NotOK: For i IN 0 to 7 GENERATE
   accum <= accum + 1 WHEN x(i)='1';
END GENERATE;
```

Example 5.6:   Vector Shifter

This example illustrates the use of GENERATE. In it, the output vector must be a shifted version of the input vector, with twice its width and an amount of shift specified by another input. For example, if the input bus has width 4, and the present value is "1111", then the output should be one of the lines of the following matrix (the original vector is underscored):

row(0): 0 0 0 0 <u>1 1 1 1</u>

row(1): 0 0 0 <u>1 1 1 1</u> 0

row(2): 0 0 <u>1 1 1 1</u> 0 0

row(3): 0 <u>1 1 1 1</u> 0 0 0

row(4): <u>1 1 1 1</u> 0 0 0 0

The first row corresponds to the input itself, with no shift and the most significant bits filled with '0's. Each successive row is equal to the previous row shifted one position to the left.

The solution below has input inp, output outp, and shift selection sel. Each row of the array above (called *matrix*, line 14) is defined as subtype vector (line 12).

```
1  --------------------------------------------------
2  LIBRARY ieee;
3  USE ieee.std_logic_1164.all;
4  --------------------------------------------------
5  ENTITY shifter IS
6     PORT ( inp: IN STD_LOGIC_VECTOR (3 DOWNTO 0);
7               sel: IN INTEGER RANGE 0 TO 4;
8               outp: OUT STD_LOGIC_VECTOR (7 DOWNTO 0));
9  END shifter;
10 --------------------------------------------------
11 ARCHITECTURE shifter OF shifter IS
12    SUBTYPE vector IS STD_LOGIC_VECTOR (7 DOWNTO 0);
13    TYPE matrix IS ARRAY (4 DOWNTO 0) OF vector;
14    SIGNAL row: matrix;
15 BEGIN
```

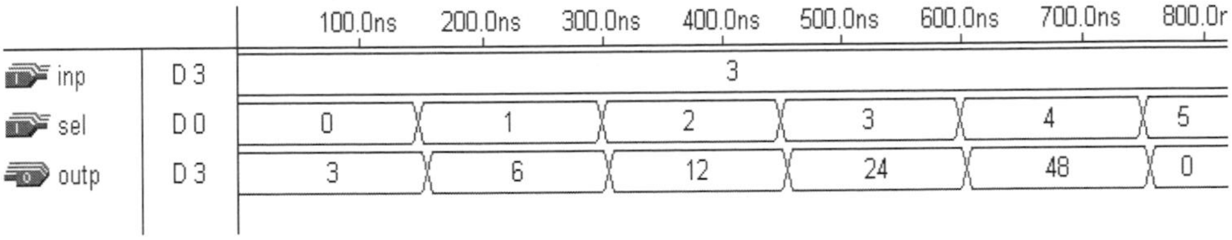

Figure 5.12
Simulation results of example 5.6.

```
16      row(0) <= "0000" & inp;
17      G1: FOR i IN 1 TO 4 GENERATE
18          row(i) <= row(i-1)(6 DOWNTO 0) & '0';
19      END GENERATE;
20      outp <= row(sel);
21 END shifter;
22 -----------------------------------------------
```

Simulation results are presented in figure 5.12. As can be seen, inp = "0011" (decimal 3) was applied to the circuit. The result was outp = "00000011" (decimal 3) when sel = 0 (no shift), outp = "00000110" (decimal 6) when sel = 1 (one shift to the left), outp = "00001100" (decimal 12) when sel = 2 (two shifts to the left), and so on.

5.5 BLOCK

There are two kinds of BLOCK statements: *Simple* and *Guarded*.

Simple BLOCK

The BLOCK statement, in its *simple* form, represents only a way of locally *partitioning* the code. It allows a set of *concurrent* statements to be *clustered* into a BLOCK, with the purpose of turning the overall code more readable and more manageable (which might be helpful when dealing with long codes). Its syntax is shown below.

```
label: BLOCK
  [declarative part]
BEGIN
  (concurrent statements)
END BLOCK label;
```

Therefore, the overall aspect of a "blocked" code is the following:

```
-----------------------
ARCHITECTURE example ...
BEGIN
    ...
    block1: BLOCK
    BEGIN
       ...
    END BLOCK block1
    ...
    block2: BLOCK
    BEGIN
       ...
    END BLOCK block2;
    ...
END example;
-----------------------
```

Example:

```
b1: BLOCK
    SIGNAL a: STD_LOGIC;
BEGIN
    a <= input_sig   WHEN ena='1' ELSE 'Z';
END BLOCK b1;
```

A BLOCK (simple or guarded) can be nested inside another BLOCK. The corresponding syntax is shown below.

```
    label1: BLOCK
      [declarative part of top block]
    BEGIN
      [concurrent statements of top block]
    label2: BLOCK
      [declarative part nested block]
    BEGIN
      (concurrent statements of nested block)
    END BLOCK label2;
      [more concurrent statements of top block]
    END BLOCK label1;
```

Note: Although code partitioning techniques are the object of Part II of the book, and the BLOCK statement seen above serves exactly to this purpose, BLOCK is described in this section due to the fact that it is self-contained within the main code (that is, it does not invoke any extra PACKAGE, COMPONENT, FUNCTION, or PROCEDURE—these four units are the actual focus of Part II).

Guarded BLOCK

A *guarded* BLOCK is a special kind of BLOCK, which includes an additional expression, called *guard* expression. A guarded statement in a guarded BLOCK is executed only when the guard expression is TRUE.

Guarded BLOCK:

```
label: BLOCK (guard expression)
  [declarative part]
BEGIN
  (concurrent guarded and unguarded statements)
END BLOCK label;
```

As the examples below illustrate, even though only *concurrent* statements can be written within a BLOCK, with a guarded BLOCK even sequential circuits can be constructed. This, however, is not a usual design approach.

Example 5.7:   Latch Implemented with a Guarded BLOCK

The example presented below implements a transparent latch. In it, clk='1' (line 12) is the guard expression, while q<=GUARDED d (line 14) is a guarded statement. Therefore, q<=d will only occur if clk='1'.

```
1  ------------------------------
2  LIBRARY ieee;
3  USE ieee.std_logic_1164.all;
4  ------------------------------
5  ENTITY latch IS
6    PORT (d, clk: IN STD_LOGIC;
7          q: OUT STD_LOGIC);
8  END latch;
9  ------------------------------
10 ARCHITECTURE latch OF latch IS
```

```
11 BEGIN
12    b1: BLOCK (clk='1')
13    BEGIN
14       q <= GUARDED d;
15    END BLOCK b1;
16 END latch;
17 -----------------------------
```

Example 5.8:   DFF Implemented with a Guarded BLOCK

Here, a positive-edge sensitive D-type flip-flop, with synchronous reset, is designed. The interpretation of the code is similar to that in the example above. In it, clk'EVENT AND clk='1' (line 12) is the guard expression, while q <= GUARDED '0' WHEN rst='1' (line 14) is a guarded statement. Therefore, q<='0' will occur when the guard expression is true *and* rst is '1'.

```
1  -----------------------------
2  LIBRARY ieee;
3  USE ieee.std_logic_1164.all;
4  -----------------------------
5  ENTITY dff IS
6     PORT ( d, clk, rst: IN STD_LOGIC;
7              q: OUT STD_LOGIC);
8  END dff;
9  -----------------------------
10 ARCHITECTURE dff OF dff IS
11 BEGIN
12    b1: BLOCK (clk'EVENT AND clk='1')
13    BEGIN
14       q <= GUARDED '0' WHEN rst='1' ELSE d;
15    END BLOCK b1;
16 END dff;
17 -----------------------------
```

5.6 Problems

The problems proposed in this section are to be solved using only *concurrent* code (operators, WHEN, GENERATE). After writing the VHDL code, synthesize and simulate it, to make sure that it works as expected.

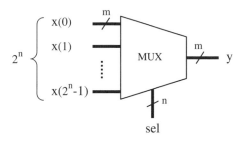

Figure P5.1

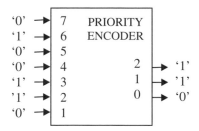

Figure P5.2

Problem 5.1:   Generic Multiplexer

We have seen the design of a multiplexer in examples 5.1 and 5.2. Those circuits were for a pre-defined number of inputs (4 inputs) and a pre-defined number of bits per input (1 bit). A truly *generic* mux is depicted in figure P5.1. In it, n represents the number of bits of the selection input (sel), while m indicates the number of bits per input. The circuit has 2^n inputs (notice that there is no relationship between m and n). Using a GENERIC statement to specify n, and assuming m = 8, design this circuit.

Suggestion: The input should be specified as an array of vectors. Therefore, review section 3.5. Does your solution (ARCHITECTURE) require more than one line of actual code?

Problem 5.2:   Priority Encoder

Figure P5.2 shows the top-level diagram of a 7-level priority encoder. The circuit must encode the address of the input bit of highest order that is active. "000" should indicate that there is no request at the input (no bit active). Write two solutions for this circuit:

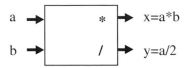

Figure P5.3

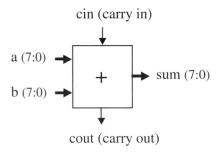

Figure P5.4

(a) Using only operators;

(b) Using WHEN/ELSE (simple WHEN);

Problem 5.3:   Simple Multiplier/Divider

Using only concurrent code, design the multiplier/divider of figure P5.3. The circuit has two 8-bit integer inputs (a, b) and two integer outputs (x, y), where x = a*b and y = a/2.

Note: For a *generic* fixed-point divider, you may consult chapter 9.

Problem 5.4:   Adder

Using only concurrent statements, design the 8-bit unsigned adder of figure P5.4.

Problem 5.5:   Signed/Unsigned Adder/Subtractor

In figure P5.5, we have added an extra 2-bit input (sel) to the circuit of problem 5.4, such that now the circuit can operate as a signed or unsigned adder/subtractor (see truth table). Write a concurrent VHDL code for this circuit.

Note: After having solved this problem, you can compare your solution to a corresponding example in chapter 9.

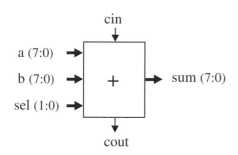

sel	operation
00	add unsigned
01	add signed
10	sub unsigned
11	sub signed

Figure P5.5

Table P5.6

Binary code	Gray code
0000	0000
0001	0001
0010	0011
0011	0010
0100	0110
0101	0111
0110	0101
0111	0100
1000	1100
1001	1101
1010	1111
1011	1110
1100	1010
1101	1011
1110	1001
1111	1000

Problem 5.6: Binary-to-Gray Code Converter

Binary code is the most often used of all digital codes. In it, the LSB (least significant bit) has weight 2^0, with the weight increasing by a factor of two for each successive bit, up to 2^{n-1} for the MSB (most significant bit), where n is the number of bits in the codeword. The Gray code, on the other hand, is based on minimum Hamming distance between neighboring codewords, that is, only one bit changes when we move from the j-th to the (j + 1)-th codeword. Both codes, for n = 4, are listed in table P5.6. Design a circuit capable of converting binary code to Gray code (for generic n). If possible, present more than one solution.

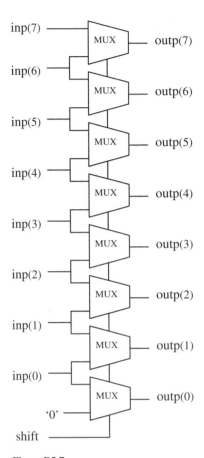

inp(7) — MUX — outp(7)

inp(6) —

MUX — outp(6)

inp(5) —

MUX — outp(5)

inp(4) —

MUX — outp(4)

inp(3) —

MUX — outp(3)

inp(2) —

MUX — outp(2)

inp(1) —

MUX — outp(1)

inp(0) —

MUX — outp(0)

'0' —

shift —

Figure P5.7

Problem 5.7:   Simple Barrel Shifter

Figure P5.7 shows the diagram of a very simple barrel shifter. In this case, the circuit must shift the input vector (of size 8) either 0 or 1 position to the left. When actually shifted (shift = 1), the LSB bit must be filled with '0' (shown in the bottom left corner of the diagram). If shift = 0, then outp = inp; else, if shift = 1, then outp(0) = '0' and outp(i) = inp(i − 1), for $1 \leq i \leq 7$. Write a concurrent code for this circuit.

Note: A complete barrel shifter (with shift = 0 to n − 1, where n is the number of bits) will be seen in chapter 9.

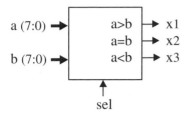

Figure P5.8

Problem 5.8: Comparator

Construct a circuit capable of comparing two 8-bit vectors, a and b. A selection pin (sel) should determine whether the comparison is signed (sel = '1') or unsigned (sel = '0'). The circuit must have three outputs, x1, x2, and x3, corresponding to a > b, a = b, and a < b, respectively (figure P5.8).

Note: After having solved this problem, you can compare your solution to a corresponding example in chapter 9.

6 Sequential Code

As mentioned in chapter 5, VHDL code is inherently *concurrent*. PROCESSES, FUNCTIONS, and PROCEDURES are the only sections of code that are executed *sequentially*. However, as a whole, any of these blocks is still concurrent with any other statements placed outside it.

One important aspect of sequential code is that it is not limited to sequential logic. Indeed, with it we can build *sequential* circuits as well as *combinational* circuits. Sequential code is also called *behavioral* code.

The statements discussed in this section are all sequential, that is, allowed only inside PROCESSES, FUNCTIONS, or PROCEDURES. They are: IF, WAIT, CASE, and LOOP.

VARIABLES are also restricted to be used in sequential code only (that is, inside a PROCESS, FUNCTION, or PROCEDURE). Thus, contrary to a SIGNAL, a VARIABLE can never be global, so its value can not be passed out directly.

We will concentrate on PROCESSES here. FUNCTIONS and PROCEDURES are very similar, but are intended for system-level design, being therefore seen in Part II of this book.

6.1 PROCESS

A PROCESS is a *sequential* section of VHDL code. It is characterized by the presence of IF, WAIT, CASE, or LOOP, and by a *sensitivity list* (except when WAIT is used). A PROCESS must be installed in the main code, and is executed every time a signal in the sensitivity list changes (or the condition related to WAIT is fulfilled). Its syntax is shown below.

```
[label:] PROCESS (sensitivity list)
  [VARIABLE name type [range] [:= initial_value;]]
BEGIN
  (sequential code)
END PROCESS [label];
```

VARIABLES are optional. If used, they must be declared in the declarative part of the PROCESS (before the word BEGIN, as indicated in the syntax above). The initial value is not synthesizable, being only taken into consideration in simulations.

The use of a label is also optional. Its purpose is to improve code readability. The label can be any word, except VHDL reserved words (appendix E).

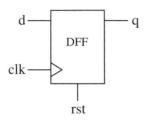

Figure 6.1
DFF with asynchronous reset of example 6.1.

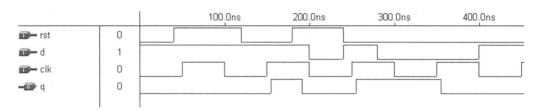

Figure 6.2
Simulation results of example 6.1.

To construct a synchronous circuit, monitoring a signal (clock, for example) is necessary. A common way of detecting a signal change is by means of the EVENT attribute (seen in section 4.2). For instance, if clk is a signal to be monitored, then clk'EVENT returns TRUE when a change on clk occurs (rising or falling edge). An example, illustrating the use of EVENT and PROCESS, is shown next.

Example 6.1:   DFF with Asynchronous Reset #1

A D-type flip-flop (DFF, figure 6.1) is the most basic building block in sequential logic circuits. In it, the output must copy the input at either the positive or negative transition of the clock signal (rising or falling edge).

In the code presented below, we make use of the IF statement (discussed in section 6.3) to design a DFF with asynchronous reset. If rst = '1', then the output must be q = '0' (lines 14–15), regardless of the status of clk. Otherwise, the output must copy the input (that is, q = d) at the positive edge of clk (lines 16–17). The EVENT attribute is used in line 16 to detect a clock transition. The PROCESS (lines 12–19) is run every time any of the signals that appear in its sensitivity list (clk and rst, line 12) changes. Simulation results, confirming the functionality of the synthesized circuit, are presented in figure 6.2.

```
1  --------------------------------------
2  LIBRARY ieee;
3  USE ieee.std_logic_1164.all;
4  --------------------------------------
5  ENTITY dff IS
6     PORT (d, clk, rst: IN STD_LOGIC;
7           q: OUT STD_LOGIC);
8  END dff;
9  --------------------------------------
10 ARCHITECTURE behavior OF dff IS
11 BEGIN
12    PROCESS (clk, rst)
13    BEGIN
14       IF (rst='1') THEN
15          q <= '0';
16       ELSIF (clk'EVENT AND clk='1') THEN
17          q <= d;
18    END IF;
19    END PROCESS;
20 END behavior;
21 --------------------------------------
```

6.2 Signals and Variables

Signals and variables will be studied in detail in the next chapter. However, it is impossible to discuss sequential code without knowing at least their most basic characteristics.

VHDL has two ways of passing non-static values around: by means of a SIGNAL or by means of a VARIABLE. A SIGNAL can be declared in a PACKAGE, ENTITY or ARCHITECTURE (in its declarative part), while a VARIABLE can only be declared inside a piece of sequential code (in a PROCESS, for example). Therefore, while the value of the former can be global, the latter is always local.

The value of a VARIABLE can never be passed out of the PROCESS directly; if necessary, then it must be assigned to a SIGNAL. On the other hand, the update of a VARIABLE is immediate, that is, we can promptly count on its new value in the next line of code. That is not the case with a SIGNAL (when used in a PROCESS), for its new value is generally only guaranteed to be available *after* the conclusion of the present run of the PROCESS.

Finally, recall from section 4.1 that the assignment operator for a SIGNAL is "<=" (ex.: sig <= 5), while for a VARIABLE it is ":=" (ex.: var := 5).

6.3 IF

As mentioned earlier, IF, WAIT, CASE, and LOOP are the statements intended for sequential code. Therefore, they can only be used inside a PROCESS, FUNCTION, or PROCEDURE.

The natural tendency is for people to use IF more than any other statement. Though this could, in principle, have a negative consequence (because the IF/ELSE statement might infer the construction of an unnecessary priority decoder), the synthesizer will optimize the structure and avoid the extra hardware. The syntax of IF is shown below.

```
IF conditions THEN assignments;
ELSIF conditions THEN assignments;
...
ELSE assignments;
END IF;
```

Example:

```
IF (x<y) THEN temp:="11111111";
ELSIF (x=y AND w='0') THEN temp:="11110000";
ELSE temp:=(OTHERS =>'0');
```

Example 6.2: One-digit Counter #1

The code below implements a progressive 1-digit decimal counter $(0 \rightarrow 9 \rightarrow 0)$. A top-level diagram of the circuit is shown in figure 6.3. It contains a single-bit input

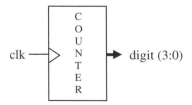

Figure 6.3
Counter of example 6.2.

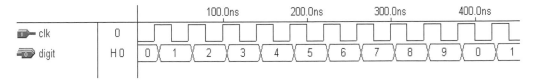

Figure 6.4
Simulation results of example 6.2.

(clk) and a 4-bit output (digit). The IF statement is used in this example. A variable, temp, was employed to create the four flip-flops necessary to store the 4-bit output signal. Simulation results, confirming the correct operation of the synthesized circuit, are shown in figure 6.4.

```
1  ---------------------------------------------
2  LIBRARY ieee;
3  USE ieee.std_logic_1164.all;
4  ---------------------------------------------
5  ENTITY counter IS
6     PORT (clk : IN STD_LOGIC;
7              digit : OUT INTEGER RANGE 0 TO 9);
8  END counter;
9  ---------------------------------------------
10 ARCHITECTURE counter OF counter IS
11 BEGIN
12    count: PROCESS(clk)
13       VARIABLE temp : INTEGER RANGE 0 TO 10;
14    BEGIN
15       IF (clk'EVENT AND clk='1') THEN
16          temp := temp + 1;
17          IF (temp=10) THEN temp := 0;
18          END IF;
19       END IF;
20       digit <= temp;
21    END PROCESS count;
22 END counter;
23 ---------------------------------------------
```

Comment: Note that the code above has neither a reset input nor any internal initialization scheme for temp (and digit, consequently). Therefore, the initial value of temp in the physical circuit can be any 4-bit value. If such value is below 10 (see line 17), the circuit will count correctly from there. On the other hand, if the value is above 10, a number of clock cycles will be used until temp reaches full count (that is, 15, or "1111"), being thus automatically reset to zero, from where the correct operation then starts. The possibility of wasting a few clock cycles in the beginning is generally not a problem. Still, if one does want to avoid that, temp = 10, in line 17, can be changed to temp <= 10, but this will increase the hardware. However, if starting exactly from 0 is always necessary, then a reset input should be included (as in example 6.7).

Notice in the code above that we increment temp and compare it to 10, with the purpose of resetting temp once 10 is reached. This is a typical approach used in counters. Notice that 10 is a constant, so a comparator to a constant is inferred by the compiler, which is a relatively simple circuit to construct. However, if instead of a constant we were using a programmable parameter, then a full comparator would need to be implemented, which requires substantially more logic than a comparator to a constant. In this case, a better solution would be to load temp with such a parameter, and then decrement it, reloading temp when the 0 value is reached. In this case, our comparator would compare temp to 0 (a constant), thus avoiding the generation of a full comparator.

Example 6.3:   Shift Register

Figure 6.5 shows a 4-bit shift register. The output bit (q) must be four positive clock edges behind the input bit (d). It also contains an asynchronous reset, which must force all flip-flop outputs to '0' when asserted. In this example, the IF statement is again employed.

```
1    -------------------------------------------------
2    LIBRARY ieee;
3    USE ieee.std_logic_1164.all;
4    -------------------------------------------------
```

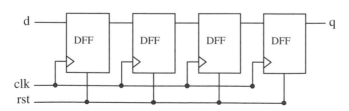

Figure 6.5
Shift register of example 6.3.

```
5   ENTITY shiftreg IS
6      GENERIC (n: INTEGER := 4);     -- # of stages
7      PORT (d, clk, rst: IN STD_LOGIC;
8            q: OUT STD_LOGIC);
9   END shiftreg;
10  ----------------------------------------------------
11  ARCHITECTURE behavior OF shiftreg IS
12     SIGNAL internal: STD_LOGIC_VECTOR (n-1 DOWNTO 0);
13  BEGIN
14     PROCESS (clk, rst)
15     BEGIN
16        IF (rst='1') THEN
17           internal <= (OTHERS => '0');
18        ELSIF (clk'EVENT AND clk='1') THEN
19           internal <= d & internal(internal'LEFT DOWNTO 1);
20        END IF;
21     END PROCESS;
22     q <= internal(0);
23  END behavior;
24  ----------------------------------------------------
```

Simulation results are shown in figure 6.6. As can be seen, q is indeed four positive clock edges behind d.

6.4 WAIT

The operation of WAIT is sometimes similar to that of IF. However, more than one form of WAIT is available. Moreover, contrary to when IF, CASE, or LOOP are

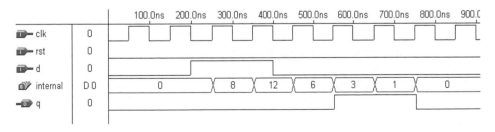

Figure 6.6
Simulation results of example 6.3.

used, the PROCESS cannot have a sensitivity list when WAIT is employed. Its syntax
(there are three forms of WAIT) is shown below.

```
WAIT UNTIL signal_condition;
```

```
WAIT ON signal1 [, signal2, ... ];
```

```
WAIT FOR time;
```

The WAIT UNTIL statement accepts only one signal, thus being more appropriate for synchronous code than asynchronous. Since the PROCESS has no sensitivity
list in this case, WAIT UNTIL must be the first statement in the PROCESS. The
PROCESS will be executed every time the condition is met.

Example: 8-bit register with synchronous reset.

```
PROCESS              -- no sensitivity list
BEGIN
   WAIT UNTIL (clk'EVENT AND clk='1');
   IF (rst='1') THEN
      output <= "00000000";
   ELSIF (clk'EVENT AND clk='1') THEN
      output <= input;
   END IF;
END PROCESS;
```

WAIT ON, on the other hand, accepts multiple signals. The PROCESS is put on
hold until any of the signals listed changes. In the example below, the PROCESS will
continue execution whenever a change in rst or clk occurs.

Example: 8-bit register with asynchronous reset.

```
PROCESS
BEGIN
   WAIT ON clk, rst;
   IF (rst='1') THEN
```

```
      output <= "00000000";
   ELSIF (clk'EVENT AND clk='1') THEN
      output <= input;
   END IF;
END PROCESS;
```

Finally, WAIT FOR is intended for simulation only (waveform generation for testbenches). Example: WAIT FOR 5ns;

Example 6.4:   DFF with Asynchronous Reset #2

The code below implements the same DFF of example 6.1 (figures 6.1 and 6.2). However, here WAIT ON is used instead of IF only.

```
1  -----------------------------------
2  LIBRARY ieee;
3  USE ieee.std_logic_1164.all;
4  -----------------------------------
5  ENTITY dff IS
6     PORT (d, clk, rst: IN STD_LOGIC;
7           q: OUT STD_LOGIC);
8  END dff;
9  -----------------------------------
10 ARCHITECTURE dff OF dff IS
11 BEGIN
12    PROCESS
13    BEGIN
14      WAIT ON rst, clk;
15      IF (rst='1') THEN
16          q <= '0';
17      ELSIF (clk'EVENT AND clk='1') THEN
18          q <= d;
19      END IF;
20    END PROCESS;
21 END dff;
22 -----------------------------------
```

Example 6.5:   One-digit Counter #2

The code below implements the same progressive 1-digit decimal counter of example 6.2 (figures 6.3 and 6.4). However, WAIT UNTIL was used instead of IF only.

```
1   ----------------------------------------------
2   LIBRARY ieee;
3   USE ieee.std_logic_1164.all;
4   ----------------------------------------------
5   ENTITY counter IS
6      PORT (clk : IN STD_LOGIC;
7              digit : OUT INTEGER RANGE 0 TO 9);
8   END counter;
9   ----------------------------------------------
10  ARCHITECTURE counter OF counter IS
11  BEGIN
12     PROCESS            -- no sensitivity list
13        VARIABLE temp : INTEGER RANGE 0 TO 10;
14     BEGIN
15        WAIT UNTIL (clk'EVENT AND clk='1');
16        temp := temp + 1;
17        IF (temp=10) THEN temp := 0;
18        END IF;
19        digit <= temp;
20     END PROCESS;
21  END counter;
22  ----------------------------------------------
```

6.5 CASE

CASE is another statement intended exclusively for sequential code (along with IF, LOOP, and WAIT). Its syntax is shown below.

```
CASE identifier IS
    WHEN value => assignments;
    WHEN value => assignments;
    ...
END CASE;
```

Example:

```
CASE control IS
   WHEN "00" => x<=a; y<=b;
```

```
    WHEN "01" => x<=b; y<=c;
    WHEN OTHERS => x<="0000"; y<="ZZZZ";
END CASE;
```

The CASE statement (sequential) is very similar to WHEN (combinational). Here too all permutations must be tested, so the keyword OTHERS is often helpful. Another important keyword is NULL (the counterpart of UNAFFECTED), which should be used when no action is to take place. For example, WHEN OTHERS => NULL;. However, CASE allows multiple assignments for each test condition (as shown in the example above), while WHEN allows only one.

Like in the case of WHEN (section 5.3), here too "WHEN value" can take up three forms:

```
WHEN value                   -- single value
WHEN value1 to value2        -- range, for enumerated data types
                             -- only
WHEN value1 | value2 |...    -- value1 or value2 or ...
```

Example 6.6:   DFF with Asynchronous Reset #3

The code below implements the same DFF of example 6.1 (figures 6.1 and 6.2). However, here CASE was used instead of IF only. Notice that a few unnecessary declarations were intentionally included in the code to illustrate their usage.

```
1  ------------------------------------------------
2  LIBRARY ieee;                    -- Unnecessary declaration,
3                                   -- because
4  USE ieee.std_logic_1164.all;   -- BIT was used instead of
5                                   -- STD_LOGIC
6  ------------------------------------------------
7  ENTITY dff IS
8     PORT (d, clk, rst: IN BIT;
9           q: OUT BIT);
10 END dff;
11 ------------------------------------------------
12 ARCHITECTURE dff3 OF dff IS
13 BEGIN
14    PROCESS (clk, rst)
15    BEGIN
16       CASE rst IS
```

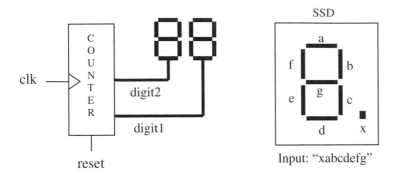

Figure 6.7
2-digit counter of example 6.7.

```
17          WHEN '1' => q<='0';
18          WHEN '0' =>
19              IF (clk'EVENT AND clk='1') THEN
20                  q <= d;
21              END IF;
22          WHEN OTHERS => NULL;    -- Unnecessary, rst is of type
23                                   -- BIT
24      END CASE;
25    END PROCESS;
26 END dff3;
27 -----------------------------------------------
```

Example 6.7: Two-digit Counter with SSD Output

The code below implements a progressive 2-digit decimal counter ($0 \rightarrow 99 \rightarrow 0$), with external asynchronous reset plus binary-coded decimal (BCD) to seven-segment display (SSD) conversion. Diagrams of the circuit and SSD are shown in figure 6.7. The CASE statement (lines 31–56) was employed to determine the output signals that will feed the SSDs. Notice that we have chosen the following connection between the circuit and the SSD: xabcdefg (that is, the MSB feeds the decimal point, while the LSB feeds segment g).

As can be seen, this circuit is a straight extension of that presented in example 6.2, with the differences that now two digits are necessary rather than one, and that the outputs must be connected to SSD displays. The operation of the circuit can be verified in the simulation results of figure 6.8.

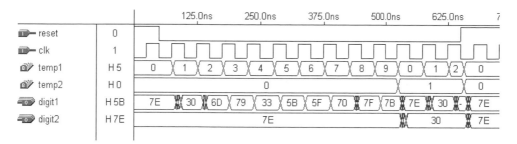

Figure 6.8
Simulation results of example 6.7.

```
1   ----------------------------------------------------
2   LIBRARY ieee;
3   USE ieee.std_logic_1164.all;
4   ----------------------------------------------------
5   ENTITY counter IS
6      PORT (clk, reset : IN STD_LOGIC;
7              digit1, digit2 : OUT STD_LOGIC_VECTOR (6 DOWNTO 0));
8   END counter;
9   ----------------------------------------------------
10  ARCHITECTURE counter OF counter IS
11  BEGIN
12     PROCESS(clk, reset)
13        VARIABLE temp1: INTEGER RANGE 0 TO 10;
14        VARIABLE temp2: INTEGER RANGE 0 TO 10;
15     BEGIN
16     ---- counter: ----------------------
17        IF (reset='1') THEN
18           temp1 := 0;
19           temp2 := 0;
20        ELSIF (clk'EVENT AND clk='1') THEN
21           temp1 := temp1 + 1;
22           IF (temp1=10) THEN
23              temp1 := 0;
24              temp2 := temp2 + 1;
25              IF (temp2=10) THEN
26                 temp2 := 0;
```

```
27              END IF;
28            END IF;
29          END IF;
30          ---- BCD to SSD conversion: --------
31          CASE temp1 IS
32             WHEN 0 => digit1 <= "1111110";    --7E
33             WHEN 1 => digit1 <= "0110000";    --30
34             WHEN 2 => digit1 <= "1101101";    --6D
35             WHEN 3 => digit1 <= "1111001";    --79
36             WHEN 4 => digit1 <= "0110011";    --33
37             WHEN 5 => digit1 <= "1011011";    --5B
38             WHEN 6 => digit1 <= "1011111";    --5F
39             WHEN 7 => digit1 <= "1110000";    --70
40             WHEN 8 => digit1 <= "1111111";    --7F
41             WHEN 9 => digit1 <= "1111011";    --7B
42             WHEN OTHERS => NULL;
43          END CASE;
44          CASE temp2 IS
45             WHEN 0 => digit2 <= "1111110";    --7E
46             WHEN 1 => digit2 <= "0110000";    --30
47             WHEN 2 => digit2 <= "1101101";    --6D
48             WHEN 3 => digit2 <= "1111001";    --79
49             WHEN 4 => digit2 <= "0110011";    --33
50             WHEN 5 => digit2 <= "1011011";    --5B
51             WHEN 6 => digit2 <= "1011111";    --5F
52             WHEN 7 => digit2 <= "1110000";    --70
53             WHEN 8 => digit2 <= "1111111";    --7F
54             WHEN 9 => digit2 <= "1111011";    --7B
55             WHEN OTHERS => NULL;
56          END CASE;
57       END PROCESS;
58 END counter;
59 ----------------------------------------------------
```

Comment: Notice above that the same routine was repeated twice (using CASE statements). We will learn, in Part II, how to write and compile frequently used pieces of code into user-defined libraries, so that such repetitions can be avoided.

6.6 LOOP

As the name says, LOOP is useful when a piece of code must be instantiated several
times. Like IF, WAIT, and CASE, LOOP is intended exclusively for sequential code,
so it too can only be used inside a PROCESS, FUNCTION, or PROCEDURE.

There are several ways of using LOOP, as shown in the syntaxes below.

FOR / LOOP: The loop is repeated a fixed number of times.

```
[label:] FOR identifier IN range LOOP
   (sequential statements)
END LOOP [label];
```

WHILE / LOOP: The loop is repeated until a condition no longer holds.

```
[label:] WHILE condition LOOP
   (sequential statements)
END LOOP [label];
```

EXIT: Used for ending the loop.

```
[label:] EXIT [label] [WHEN condition];
```

NEXT: Used for skipping loop steps.

```
[label:] NEXT [loop_label] [WHEN condition];
```

Example of FOR / LOOP:

```
FOR i IN 0 TO 5 LOOP
   x(i) <= enable AND w(i+2);
   y(0, i) <= w(i);
END LOOP;
```

In the code above, the loop will be repeated unconditionally until i reaches 5 (that
is, six times).

One important remark regarding FOR / LOOP (similar to that made for GEN-ERATE, in chapter 5) is that both limits of the range must be static. Thus a declaration of the type "FOR i IN 0 TO choice LOOP", where choice is an input (non-static) parameter, is generally not synthesizable.

Example of WHILE / LOOP: In this example, LOOP will keep repeating while i < 10.

```
WHILE (i < 10) LOOP
   WAIT UNTIL clk'EVENT AND clk='1';
   (other statements)
END LOOP;
```

Example with EXIT: In the code below, EXIT implies not an escape from the current iteration of the loop, but rather a definite exit (that is, even if i is still within the data range, the LOOP statement will be considered as concluded). In this case, the loop will end as soon as a value different from '0' is found in the data vector.

```
FOR i IN data'RANGE LOOP
   CASE data(i) IS
      WHEN '0' => count:=count+1;
      WHEN OTHERS => EXIT;
   END CASE;
END LOOP;
```

Example with NEXT: In the example below, NEXT causes LOOP to skip one iteration when i = skip.

```
FOR i IN 0 TO 15 LOOP
   NEXT WHEN i=skip;      -- jumps to next iteration
      (...)
END LOOP;
```

Several complete design examples, illustrating various applications of LOOP, are presented below.

Example 6.8:   Carry Ripple Adder

Figure 6.9 shows an 8-bit unsigned carry ripple adder. The top-level diagram shows the inputs and outputs of the circuit: a and b are the input vectors to be added, cin is the carry-in bit, s is the sum vector, and cout is the carry-out bit. The one-level-below-top diagram shows how the carry bits propagate (ripple).

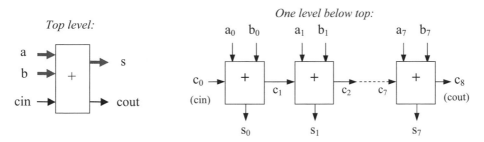

Figure 6.9
8-bit carry-ripple adder of example 6.8

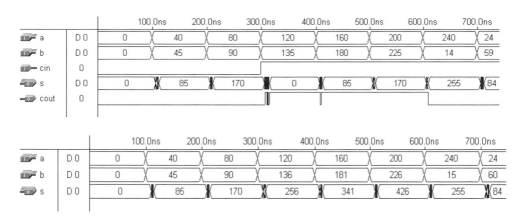

Figure 6.10
Simulation results of example 6.8.

Each section of the latter diagram is a full-adder unit (section 1.4). Thus its outputs can be computed by means of:

$s_j = a_j$ XOR b_j XOR c_j

$c_{j+1} = (a_j$ AND $b_j)$ OR $(a_j$ AND $c_j)$ OR $(b_j$ AND $c_j)$

Two *generic* solutions are presented below, one with logic vectors and the other with integers, being the simulation results for both solutions displayed in figure 6.10.

Notice, however, that there is a major difference between these two solutions. While solution 1 is a structural design (that is, we are telling the compiler the exact structure that we want for our circuit—a carry-ripple adder in this case), in solution

2 we are leaving that choice to the compiler. Indeed, when using programmable logic devices (CPLDs or FPGAs, appendix A), pure carry-ripple adders will generally not result, because the logic cells are prepared for optimally implementing faster adders at minimum hardware cost (using built-in fast carry chains, for example).

In conclusion, though the carry-ripple adder architecture is the most economical adder structure in terms of hardware (at the expense of speed) when the design is done from scratch (at the transistor level, like in an ASIC), this is not necessarily true when implemented in a PLD. In the latter, a bigger (and slower) circuit might result (see problem 6.18). Therefore, the use of solution 1 is only recommended when the VHDL code is intended for an ASIC and minimum circuit size is desired. For PLDs, the use of the "+" operator (as in solution 2) is recommended.

Note: We will see more about adders in chapter 9.

```
1  ----- Solution 1: Generic, with VECTORS --------
2  LIBRARY ieee;
3  USE ieee.std_logic_1164.all;
4  ------------------------------------------------
5  ENTITY adder IS
6     GENERIC (length : INTEGER := 8);
7     PORT ( a, b: IN STD_LOGIC_VECTOR (length-1 DOWNTO 0);
8            cin: IN STD_LOGIC;
9            s: OUT STD_LOGIC_VECTOR (length-1 DOWNTO 0);
10           cout: OUT STD_LOGIC);
11 END adder;
12 ------------------------------------------------
13 ARCHITECTURE adder OF adder IS
14 BEGIN
15    PROCESS (a, b, cin)
16       VARIABLE carry : STD_LOGIC_VECTOR (length DOWNTO 0);
17    BEGIN
18       carry(0) := cin;
19       FOR i IN 0 TO length-1 LOOP
20          s(i) <= a(i) XOR b(i) XOR carry(i);
21          carry(i+1) := (a(i) AND b(i)) OR (a(i) AND
22                          carry(i)) OR (b(i) AND carry(i));
23       END LOOP;
24       cout <= carry(length);
```

```
25     END PROCESS;
26 END adder;
27 -------------------------------------------------

1  ---- Solution 2: generic, with INTEGERS ----
2 ENTITY adder IS
3    GENERIC (bits: INTEGER := 7)
4    PORT (a, b: IN INTEGER RANGE 0 TO 2**bits-1;
              s: OUT INTEGER RANGE 0 TO 2** (bits+1)-1);
5 END adder
6  -------------------------------------------------
7 ARCHITECTURE adder OF adder IS
8 BEGIN
9    s <= a+b;
10 END adder;
11 -------------------------------------------------
```

Example 6.9:   Simple Barrel Shifter

Figure 6.11 shows the diagram of a very simple barrel shifter. In this case, the circuit must shift the input vector (of size 8) either 0 or 1 position to the left. When actually shifted (shift = 1), the LSB bit must be filled with '0' (shown in the botton left corner of the diagram). If shift = 0, then outp = inp; if shift = 1, then outp(0) = '0' and outp(i) = inp(i − 1), for $1 \le i \le 7$.

A complete VHDL code is presented below, which illustrates the use of FOR/ LOOP. Simulation results appear in figure 6.12.

Note: A complete barrel shifter (with shift = 0 to n − 1, where n is the size of the input vector) will be seen in chapter 9.

```
1  -------------------------------------------------
2  LIBRARY ieee;
3  USE ieee.std_logic_1164.all;
4  -------------------------------------------------
5  ENTITY barrel IS
6     GENERIC (n: INTEGER := 8);
7     PORT ( inp: IN STD_LOGIC_VECTOR (n-1 DOWNTO 0);
              shift: IN INTEGER RANGE 0 TO 1;
              outp: OUT STD_LOGIC_VECTOR (n-1 DOWNTO 0));
```

inp(7)
inp(6)
inp(5)
inp(4)
inp(3)
inp(2)
inp(1)
inp(0)
'0'
shift
MUX
outp(7)
outp(6)
outp(5)
outp(4)
outp(3)
outp(2)
outp(1)
outp(0)

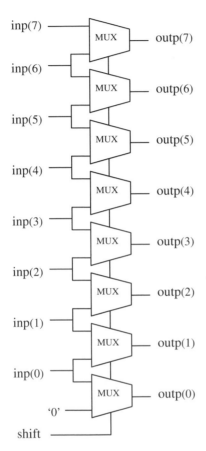

Figure 6.11
Simple barrel shifter of example 6.9.

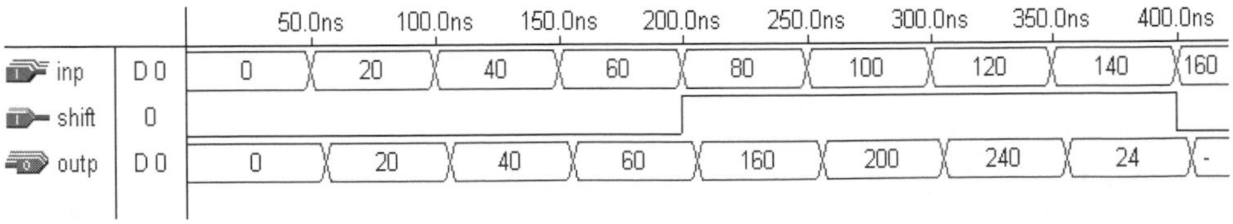

50.0ns 100.0ns 150.0ns 200.0ns 250.0ns 300.0ns 350.0ns 400.0ns
inp D 0 0 20 40 60 80 100 120 140 160
shift 0
outp D 0 0 20 40 60 160 200 240 24

Figure 6.12
Simulation results of example 6.9.

```
10 END barrel;
11 ---------------------------------------------
12 ARCHITECTURE RTL OF barrel IS
13 BEGIN
14    PROCESS (inp, shift)
15    BEGIN
16       IF (shift=0) THEN
17          outp <= inp;
18       ELSE
19          outp(0) <= '0';
20          FOR i IN 1 TO inp'HIGH LOOP
21             outp(i) <= inp(i-1);
22          END LOOP;
23       END IF;
24    END PROCESS;
25 END RTL;
26 ---------------------------------------------
```

Example 6.10:   Leading Zeros

The design below counts the number of leading zeros in a binary vector, starting from the left end. The solution illustrates the use of LOOP / EXIT. Recall that EXIT implies not a escape from the current iteration of the loop, but rather a definite exit from it (that is, even if i is still within the specified range, the LOOP statement will be considered as concluded). In this example, the loop will end as soon as a '1' is found in the data vector. Therefore, it is appropriate for counting the number of zeros that precedes the first one.

```
1  ---------------------------------------------
2  LIBRARY ieee;
3  USE ieee.std_logic_1164.all;
4  ---------------------------------------------
5  ENTITY LeadingZeros IS
6     PORT ( data: IN STD_LOGIC_VECTOR (7 DOWNTO 0);
7              zeros: OUT INTEGER RANGE 0 TO 8);
8  END LeadingZeros;
9  ---------------------------------------------
10 ARCHITECTURE behavior OF LeadingZeros IS
```

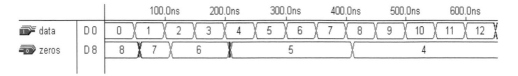

Figure 6.13
Simulation results of example 6.10.

```
11 BEGIN
12    PROCESS (data)
13       VARIABLE count: INTEGER RANGE 0 TO 8;
14    BEGIN
15       count := 0;
16       FOR i IN data'RANGE LOOP
17          CASE data(i) IS
18             WHEN '0' => count := count + 1;
19             WHEN OTHERS => EXIT;
20          END CASE;
21       END LOOP;
22       zeros <= count;
23    END PROCESS;
24 END behavior;
25 ---------------------------------------------
```

Simulation results, verifying the functionality of the circuit, are shown in figure 6.13. With data = "00000000" (decimal 0), eight zeros are detected; when data = "00000001" (decimal 1), seven zeros are encountered; etc.

6.7 CASE versus IF

Though in principle the presence of ELSE in the IF/ELSE statement might infer the implementation of a priority decoder (which would never occur with CASE), this will generally not happen. For instance, when IF (a *sequential* statement) is used to implement a fully *combinational* circuit, a multiplexer might be inferred instead. Therefore, after optimization, the general tendency is for a circuit synthesized from a VHDL code based on IF not to differ from that based on CASE.

Table 6.1
Comparison between WHEN and CASE.

	WHEN	CASE
Statement type	Concurrent	Sequential
Usage	Only outside PROCESSES, FUNCTIONS, or PROCEDURES	Only inside PROCESSES, FUNCTIONS, or PROCEDURES
All permutations must be tested	Yes for WITH/SELECT/WHEN	Yes
Max. # of assignments per test	1	Any
No-action keyword	UNAFFECTED	NULL

Example: The codes below implement the same physical multiplexer circuit.

```
---- With IF: --------------
IF (sel="00") THEN x<=a;
ELSIF (sel="01") THEN x<=b;
ELSIF (sel="10") THEN x<=c;
ELSE x<=d;

---- With CASE: ------------
CASE sel IS
    WHEN "00" => x<=a;
    WHEN "01" => x<=b;
    WHEN "10" => x<=c;
    WHEN OTHERS => x<=d;
END CASE;
---------------------------
```

6.8   CASE versus WHEN

CASE and WHEN are very similar. However, while one is concurrent (WHEN), the other is sequential (CASE). Their main similarities and differences are summarized in table 6.1.

Example: From a functional point of view, the two codes below are equivalent.

```
---- With WHEN: ----------------
WITH sel SELECT
```

```
x <=    a WHEN "000",
        b WHEN "001",
        c WHEN "010",
        UNAFFECTED WHEN OTHERS;

---- With CASE: ----------------
CASE sel IS
   WHEN "000" => x<=a;
   WHEN "001" => x<=b;
   WHEN "010" => x<=c;
   WHEN OTHERS => NULL;
END CASE;
-------------------------------
```

6.9 Bad Clocking

The compiler will generally not be able to synthesize codes that contain assignments to the same signal at both transitions of the reference (clock) signal (that is, at the rising edge plus at the falling edge). This is particularly true when the target technology contains only single-edge flip-flops (CPLDs, for example—appendix A). In this case, the compiler might display a message of the type "signal does not hold value after clock edge" or similar.

As an example, let us consider the case of a counter that must be incremented at every clock transition (rising plus falling edge). One alternative could be the following:

```
PROCESS (clk)
BEGIN
   IF(clk'EVENT AND clk='1') THEN
      counter <= counter + 1;
   ELSIF(clk'EVENT AND clk='0') THEN
      counter <= counter + 1;
   END IF;
   ...
END PROCESS;
```

In this case, besides the messages already described, the compiler might also complain that the signal `counter` is multiply driven. In any case, compilation will be suspended.

Another important aspect is that the EVENT attribute must be related to a test condition. For example, the statement IF(clk'EVENT AND clk='1') is correct, but using simply IF(clk'EVENT) will either have the compiler assume a default test value (say "AND clk='1'") or issue a message of the type "clock not locally stable". As an example, let us consider again the case of a counter that must be incremented at both transitions of clk. One could write:

```
PROCESS (clk)
BEGIN
   IF(clk'EVENT) THEN
      counter := counter + 1;
   END IF;
   ...
END PROCESS;
```

Since the **PROCESS** above is supposed to be run every time clk changes, one might expect the counter to be incremented twice per clock cycle. However, for the reason already mentioned, this will not happen. If the compiler assumes a default value, a wrong circuit will be synthesized, because only one edge of clk will be considered; if no default value is assumed, then an error message and no compilation should be expected.

Finally, if a signal appears in the sensitivity list, but does not appear in any of the assignments that compose the **PROCESS**, then it is likely that the compiler will simply ignore it. This fact can be illustrated with the double-edge counter described above once again. Say that the following code is used:

```
PROCESS (clk)
BEGIN
   counter := counter + 1;
   ...
END PROCESS;
```

This code reinforces the desire that the signal counter be incremented whenever an event occurs on clk (rising plus falling edge). However, a message of the type "ignored unnecessary pin clk" might be issued instead.

Example: Contrary to the cases described above, the 2-process code shown below will be correctly synthesized by any compiler. However, notice that we have used a different signal in each process.

```
----------------------
PROCESS (clk)
BEGIN
   IF(clk'EVENT AND clk='1') THEN
      x <= d;
   END IF;
END PROCESS;
----------------------
PROCESS (clk)
BEGIN
   IF(clk'EVENT AND clk='0') THEN
      y <= d;
   END IF;
END PROCESS;
----------------------
```

Now that you know what you can and what you should not to do, you are invited to solve problem 6.1.

Example 6.11:   RAM

Below is another example using sequential code, particularly the IF statement. We show the implementation of a RAM (random access memory).

As can be seen in figure 6.14(a), the circuit has a data input bus (data_in), a data output bus (data_out), an address bus (addr), plus clock (clk) and write enable

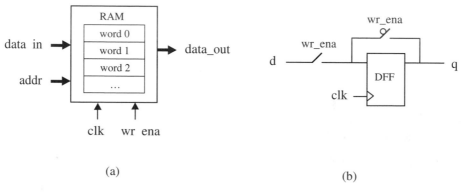

(a)   (b)

Figure 6.14
RAM circuit of example 6.11.

(wr_ena) pins. When wr_ena is asserted, at the next rising edge of clk the vector present at data_in must be stored in the position specified by addr. The output, data_out, on the other hand, must constantly display the data selected by addr.

From the register point-of-view, the circuit can be summarized as in figure 6.14(b). When wr_ena is low, q is connected to the input of the flip-flop, and terminal d is open, so no new data will be written into the memory. However, when wr_ena is turned high, d is connected to the input of the register, so at the next rising edge of clk d will overwrite its previous value.

A VHDL code that implements the circuit of figure 6.14 is shown below. The capacity chosen for the RAM is 16 words of length 8 bits each. Notice that the code is totally generic.

Note: Other memory implementations will be presented in section 9.10 of chapter 9.

```
1   -----------------------------------------------------
2   LIBRARY ieee;
3   USE ieee.std_logic_1164.all;
4   -----------------------------------------------------
5   ENTITY ram IS
6   GENERIC ( bits: INTEGER := 8;      -- # of bits per word
7             words: INTEGER := 16);   -- # of words in the memory
8      PORT ( wr_ena, clk: IN STD_LOGIC;
9             addr: IN INTEGER RANGE 0 TO words-1;
10            data_in: IN STD_LOGIC_VECTOR (bits-1 DOWNTO 0);
11            data_out: OUT STD_LOGIC_VECTOR (bits-1 DOWNTO 0));
12  END ram;
13  -----------------------------------------------------
14  ARCHITECTURE ram OF ram IS
15     TYPE vector_array IS ARRAY (0 TO words-1) OF
16        STD_LOGIC_VECTOR (bits-1 DOWNTO 0);
17     SIGNAL memory: vector_array;
18  BEGIN
19     PROCESS (clk, wr_ena)
20     BEGIN
21        IF (wr_ena='1') THEN
22           IF (clk'EVENT AND clk='1') THEN
23              memory(addr) <= data_in;
24           END IF;
25        END IF;
```

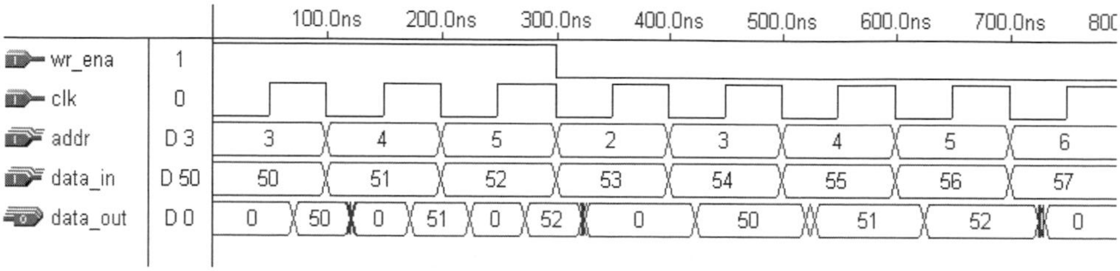

Figure 6.15
Simulation results of example 6.11.

```
26     END PROCESS;
27     data_out <= memory(addr);
28 END ram;
29 --------------------------------------------------
```

Simulation results from the circuit synthesizad with the code above are shown in figure 6.15.

6.10   Using Sequential Code to Design Combinational Circuits

We have already seen that sequential code can be used to implement either sequential or combinational circuits. In the former case, registers are necessary, so will be inferred by the compiler. However, this should not happen in the latter case. Moreover, if the code is intended for a combinational circuit, then the complete truth-table should be clearly specified in the code.

In order to satisfy the criteria above, the following rules should be observed:

Rule 1: Make sure that all input signals used (read) in the PROCESS appear in its sensitivity list.

Rule 2: Make sure that all combinations of the input/output signals are included in the code; that is, make sure that, by looking at the code, the circuit's complete truth-table can be obtained (indeed, this is true for both sequential as well as concurrent code).

Failing to comply with rule 1 will generally cause the compiler to simply issue a warning saying that a given input signal was not included in the sensitivity list, and then proceed as if the signal were included. Even though no damage is caused to the design in this case, it is a good design practice to always take rule 1 into consideration.

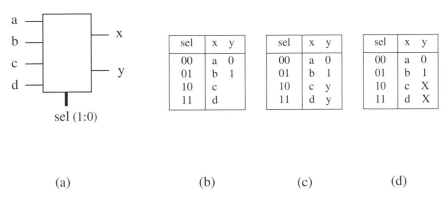

Figure 6.16
Circuit of example 6.12: (a) top-level diagram, (b) specifications provided, (c) implemented truth-table, and (d) the right approach.

With respect to rule 2, however, the consequences can be more serious because incomplete specifications of the output signals might cause the synthesizer to infer latches in order to hold their previous values. This fact is illustrated in the example below.

Example 6.12:   Bad Combinational Design

Let us consider the circuit of figure 6.16, for which the following specifications have been provided: x should behave as a multiplexer; that is, should be equal to the input selected by sel; y, on the other hand, should be equal to '0' when sel = "00", or '1' if sel = "01". These specifications are summarized in the truth-table of figure 6.16(b).

Notice that this is a combinational circuit. However, the specifications provided for y are incomplete, as can be observed in the truth-table of figure 6.16(b). Using just these specifications, the code could be the following:

```
1   ------------------------------------
2   LIBRARY ieee;
3   USE ieee.std_logic_1164.all;
4   ------------------------------------
5   ENTITY example IS
6      PORT (a, b, c, d: IN STD_LOGIC;
7           sel: IN INTEGER RANGE 0 TO 3;
8           x, y: OUT STD_LOGIC);
9   END example;
10  ------------------------------------
```

```
11 ARCHITECTURE example OF example IS
12 BEGIN
13     PROCESS (a, b, c, d, sel)
14     BEGIN
15         IF (sel=0) THEN
16             x<=a;
17             y<='0';
18         ELSIF (sel=1) THEN
19             x<=b;
20             y<='1';
21         ELSIF (sel=2) THEN
22             x<=c;
23         ELSE
24             x<=d;
25         END IF;
26     END PROCESS;
27 END example;
28 -------------------------------------
```

After compiling this code, the report files show that no flip-flops were inferred (as expected). However, when we look at the simulation results (figure 6.17), we notice something peculiar about y. Observe that, for the same value of the input (sel = 3 = "11"), two different results are obtained for y (when sel = 3 is preceded by sel = 0, y = '0' results, while y = '1' is obtained when sel = 3 is preceded by sel = 1). This signifies that some sort of memory was indeed implemented by the compiler. In fact, if we look at the equations obtained with Quartus II, for example (appendix D), we verify that y was computed as y = (sel(0) AND sel(1)) OR (sel(0) AND y) OR

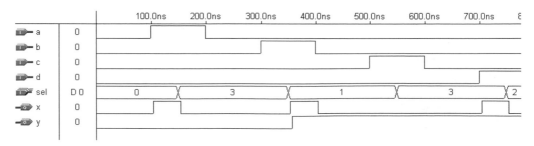

Figure 6.17
Simulation results of example 6.12.

(sel(1) AND y). Therefore, a latch (using AND/OR gates) was implemented, which renders the truth-table of figure 6.16(c).

To avoid the extra logic required by the latch, the specifications of figure 6.16(d) should be used ('X' was used for all unknown or "don't care" values). Thus the line y<='X'; must be included below lines 22 and 24 in the code above. Now, y can be as simple as y = sel(0).

6.11 Problems

Like the examples just seen, the purpose of the problems proposed in this section is to further illustrate the construction of *sequential* code (that is, the use of IF, WAIT, CASE, and LOOP, always inside a PROCESS). However, if you want to know more about SIGNALS and VARIABLES before working on the problems below, you may have a look at chapter 7, and then return to this section. Finally, recall that with sequential code we can implement sequential as well as combinational logic circuits. Though you will be using only sequential code in this section, you are invited to determine whether each circuit in the problems below (and in the examples just seen, for that matter) is actually a combinational or sequential circuit.

Problem 6.1:   Event Counter

Design a circuit capable of counting the number of clock events (number of rising edges + falling edges, figure P6.1).

Problem 6.2:   Shift Register

Write a VHDL code that implements the 4-stage shift-register of figure P6.2. The solution should be different from that of example 6.3.

Problem 6.3:   Priority Encoder

Figure P6.3 shows the same priority encoder of problem 5.2. The circuit must encode the address of the input bit of highest order that is active. The output "000" should indicate that there is no request at the input (no bit active). Write a VHDL solution for this circuit using only *sequential* code. Present two solutions:

clk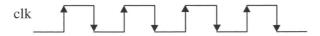

Figure P6.1

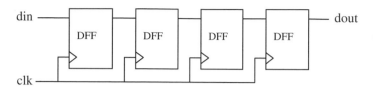

Figure P6.2

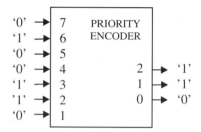

Figure P6.3

Figure P6.4

(a) With IF.

(b) With CASE.

Problem 6.4:   Generic Frequency Divider

Write a VHDL code for a circuit capable of dividing the frequency of an input clock signal by an integer n (figure P6.4). The code should be generic; that is, n should be defined using the GENERIC statement.

Problem 6.5:   Frequency Multiplier

What about the opposite of problem 6.4, that is, say that we want to *multiply* the clock frequency by n. Can it be done?

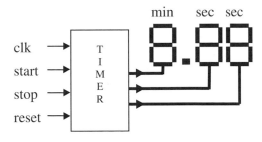

Figure P6.6

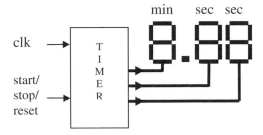

Figure P6.7

Problem 6.6: Timer #1

Design a timer capable of running from 0min:00sec to 9min:59sec (figure P6.6). The circuit must have start, stop, and reset buttons. The outputs must be SSD coded. Consider that a reliable 1 Hz clock signal is available.

Problem 6.7: Timer #2

Consider the timer of problem 6.6. However, say that now only one button is available, which must perform the start and stop functions alternately, and it also resets the circuit when pressed for more than 2 seconds. Write a VHDL code for such a timer (figure P6.7). Again, consider that a reliable 1 Hz clock is available.

Problem 6.8: Parity Detector

Figure P6.8 shows the top-level diagram of a parity detector. The input vector has eight bits. The output must be '0' when the number of '1's in the input vector is even, or '1' otherwise. Write a sequential code for this circuit. If possible, write more than one solution.

Figure P6.8

Table P6.9

Number of ones in din(7:1)	count(2:0)
0	000
1	001
2	010
3	011
4	100
5	101
6	110
7	111

Table P6.10

Number of ones in din(7:1)	dout(7:0)
0	00000001
1	00000010
2	00000100
3	00001000
4	00010000
5	00100000
6	01000000
7	10000000

Problem 6.9:   Count Ones

Say that we want to design a circuit that counts the number of '1's in a given binary vector (table P6.9). Write a VHDL code that implements such a circuit. Then synthesize and test your solution.

Problem 6.10:   Intensity Encoder

Design an encoder that receives as input a 7-bit vector din, and creates from it an output vector dout whose bits are all '0's, except the bit whose index corresponds to the number of '1's in din. All possible situations are summarized in table P6.10.

Problem 6.11: Multiplexer

Write a *sequential* VHDL code for the circuit of problem 5.1. If possible, present more than one solution.

Problem 6.12: Vector Shifter

Write a *sequential* VHDL code for the circuit of example 5.6. If possible, present more than one solution.

Problem 6.13: ALU

Write a *sequential* VHDL code for the circuit of example 5.5. If possible, present more than one solution.

Problem 6.14: Signed/Unsigned Adder/Subtractor

Solve problem 5.5 using *sequential* code. Make the code as *generic* as possible.

Problem 6.15: Comparator

Solve problem 5.8 using *sequential* code.

Problem 6.16: Carry Ripple Adder

Consider the carry ripple adder of example 6.8.

(a) Why cannot we replace the IF statement of lines 17–19 in solution 2 by simply "`temp:=c0;`"?

(b) Notice that the circuit of example 6.8 is fully combinational, so it can also be implemented using only concurrent code (that is, without a PROCESS). Write such a code for it. Then simulate it and analyze the results.

Problem 6.17: DFF

Consider the DFF with asynchronous reset of figure 6.1. Below are several codes for that circuit. Examine each of them and determine whether they should work properly. Briefly explain your answers.

```
---------------------------------------
LIBRARY ieee;
USE ieee.std_logic_1164.all;
---------------------------------------
ENTITY dff IS
```

```
   PORT ( d, clk, rst: IN BIT;
         q: OUT BIT);
END dff;
----- Solution 1 --------------------
ARCHITECTURE arch1 OF dff IS
BEGIN
   PROCESS (clk, rst)
   BEGIN
      IF (rst='1') THEN
         q <= '0';
      ELSIF (clk'EVENT AND clk='1') THEN
         q <= d;
      END IF;
   END PROCESS;
END arch1;
----- Solution 2 --------------------
ARCHITECTURE arch2 OF dff IS
BEGIN
   PROCESS (clk)
   BEGIN
      IF (rst='1') THEN
         q <= '0';
      ELSIF (clk'EVENT AND clk='1') THEN
         q <= d;
      END IF;
   END PROCESS;
END arch2;
----- Solution 3 --------------------
ARCHITECTURE arch3 OF dff IS
BEGIN
   PROCESS (clk)
   BEGIN
      IF (rst='1') THEN
         q <= '0';
      ELSIF (clk'EVENT) THEN
         q <= d;
      END IF;
   END PROCESS;
END arch3;
```

```
----- Solution 4 --------------------
ARCHITECTURE arch4 OF dff IS
BEGIN
   PROCESS (clk)
   BEGIN
      IF (rst='1') THEN
         q <= '0';
      ELSIF (clk='1') THEN
         q <= d;
      END IF;
   END PROCESS;
END arch4;
----- Solution 5 --------------------
ARCHITECTURE arch5 OF dff IS
BEGIN
   PROCESS (clk, rst, d)
   BEGIN
      IF (rst='1') THEN
         q <= '0';
      ELSIF (clk='1') THEN
         q <= d;
      END IF;
   END PROCESS;
END arch5;
---------------------------------------
```

Problem 6.18:   Adders

Compile both solutions of example 6.8 and compare the following:

a) Number of logic cells required
b) Maximum delay

Do it for at least one CPLD and one FPGA.

7 Signals and Variables

VHDL provides two objects for dealing with non-static data values: SIGNAL and VARIABLE. It also provides means for establishing default (static) values: CONSTANT and GENERIC. The last of these (the GENERIC attribute) was already seen in chapter 4. SIGNAL, VARIABLE, and CONSTANT will be studied together in this chapter.

CONSTANT and SIGNAL can be *global* (that is, seen by the whole code), and can be used in either type of code, concurrent or sequential. A VARIABLE, on the other hand, is *local*, for it can only be used inside a piece of sequential code (that is, in a PROCESS, FUNCTION, or PROCEDURE) and its value can never be passed out directly.

As will become apparent, the choice between a SIGNAL or a VARIABLE is not always easy, so an entire section and several examples will be devoted to the matter. Moreover, a discussion on the number of registers inferred by the compiler, based on SIGNAL and VARIABLE assignments, will also be presented.

7.1 CONSTANT

CONSTANT serves to establish default values. Its syntax is shown below.

```
CONSTANT name : type := value;
```

Examples:

```
CONSTANT set_bit : BIT := '1';
CONSTANT datamemory : memory := (('0','0','0','0'),
                                 ('0','0','0','1'),
                                 ('0','0','1','1'));
```

A CONSTANT can be declared in a PACKAGE, ENTITY, or ARCHITECTURE. When declared in a package, it is truly global, for the package can be used by several entities. When declared in an entity (after PORT), it is global to all architectures that follow that entity. Finally, when declared in an architecture (in its declarative part), it is global only to that architecture's code. The most common places to find a CONSTANT declaration is in an ARCHITECTURE or in a PACKAGE.

7.2 SIGNAL

SIGNAL serves to pass values in and out the circuit, as well as between its internal units. In other words, a signal represents circuit interconnects (wires). For instance, all PORTS of an ENTITY are signals by default. Its syntax is the following:

```
SIGNAL name : type [range] [:= initial_value];
```

Examples:

```
SIGNAL control: BIT := '0';
SIGNAL count: INTEGER RANGE 0 TO 100;
SIGNAL y: STD_LOGIC_VECTOR (7 DOWNTO 0);
```

The declaration of a SIGNAL can be made in the same places as the declaration of a CONSTANT (described above).

A very important aspect of a SIGNAL, when used inside a section of sequential code (PROCESS, for example), is that its update is *not* immediate. In other words, its new value should not be expected to be ready before the conclusion of the corresponding PROCESS, FUNCTION or PROCEDURE.

Recall that the assignment operator for a SIGNAL is "<=" (Ex.: count<=35;). Also, the initial value in the syntax above is not synthesizable, being only considered in simulations.

Another aspect that might affect the result is when multiple assignments are made to the same SIGNAL. The compiler might complain and quit synthesis, or might infer the wrong circuit (by considering only the last assignment, for example). Therefore, establishing initial values, like in line 15 of the example below, should be done with a VARIABLE.

Example 7.1:   Count Ones #1 (not OK)

Say that we want to design a circuit that counts the number of '1's in a binary vector (problem 6.9). Let us consider the solution below, which uses only signals. This code has multiple assignments to the same signal, temp, in lines 15 (once) and 18 (eight times). Moreover, since the value of a signal is not updated immediately, line 18 conflicts with line 15, for the value assigned in line 15 might not be ready until the conclusion of the PROCESS, in which case a wrong value would be computed in line 18. In this kind of situation, the use of a VARIABLE is recommended (example 7.2).

```
1  ----------------------------------------
2  LIBRARY ieee;
3  USE ieee.std_logic_1164.all;
4  ----------------------------------------
5  ENTITY count_ones IS
6     PORT ( din: IN STD_LOGIC_VECTOR (7 DOWNTO 0);
7              ones: OUT INTEGER RANGE 0 TO 8);
8  END count_ones;
9  ----------------------------------------
10 ARCHITECTURE not_ok OF count_ones IS
11    SIGNAL temp: INTEGER RANGE 0 TO 8;
12 BEGIN
13    PROCESS (din)
14    BEGIN
15       temp <= 0;
16       FOR i IN 0 TO 7 LOOP
17          IF (din(i)='1') THEN
18             temp <= temp + 1;
19          END IF;
20       END LOOP;
21       ones <= temp;
22    END PROCESS;
23 END not_ok;
24 ----------------------------------------
```

Notice also in the solution above that the internal signal temp (line 11) seems unnecessary, because ones could have been used directly. However, to do so, the mode of ones would need to be changed from OUT to BUFFER (line 7), because ones is assigned a value *and* is also read (used) internally. Nevertheless, since ones is a genuine unidirectional (OUT) signal, the use of an auxiliary signal (temp) is an adequate design practice.

7.3 VARIABLE

Contrary to CONSTANT and SIGNAL, a VARIABLE represents only local information. It can only be used inside a PROCESS, FUNCTION, or PROCEDURE (that is, in sequential code), and its value can not be passed out directly. On the other hand, its update is immediate, so the new value can be promptly used in the next line of code.

To declare a VARIABLE, the following syntax should be used:

```
VARIABLE name : type [range] [:= init_value];
```

Examples:

```
VARIABLE control: BIT := '0';
VARIABLE count: INTEGER RANGE 0 TO 100;
VARIABLE y: STD_LOGIC_VECTOR (7 DOWNTO 0) := "10001000";
```

Since a VARIABLE can only be used in sequential code, its declaration can only be done in the declarative part of a PROCESS, FUNCTION, or PROCEDURE.

Recall that the assignment operator for a VARIABLE is ":=" (Ex.: count:=35;). Also, like in the case of a SIGNAL, the initial value in the syntax above is not synthesizable, being only considered in simulations.

Example 7.2:   Count Ones #2 (OK)

Let us consider the problem of example 7.1 once again. The only difference in the solution below is that an internal VARIABLE is employed instead of a SIGNAL. Since the update of a variable is immediate, the initial value is established correctly and no complains regarding multiple assignments will be issued by the compiler. Simulation results can be verified in figure 7.1.

```
1   ----------------------------------------
2   LIBRARY ieee;
3   USE ieee.std_logic_1164.all;
4   ----------------------------------------
5   ENTITY count_ones IS
6       PORT ( din: IN STD_LOGIC_VECTOR (7 DOWNTO 0);
7                ones: OUT INTEGER RANGE 0 TO 8);
```

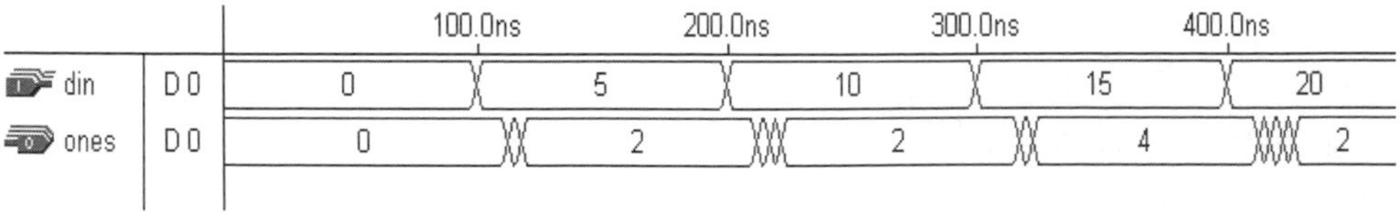

Figure 7.1
Simulation results of example 7.2.

```
8   END count_ones;
9   ----------------------------------------
10  ARCHITECTURE ok OF count_ones IS
11  BEGIN
12     PROCESS (din)
13        VARIABLE temp: INTEGER RANGE 0 TO 8;
14     BEGIN
15        temp := 0;
16        FOR i IN 0 TO 7 LOOP
17           IF (din(i)='1') THEN
18              temp := temp + 1;
19           END IF;
20        END LOOP;
21        ones <= temp;
22     END PROCESS;
23  END ok;
24  ----------------------------------------
```

7.4 SIGNAL versus VARIABLE

As already mentioned, choosing between a SIGNAL or a VARIABLE is not always straightforward. Their main differences are summarized in table 7.1.

Table 7.1
Comparison between SIGNAL and VARIABLE.

	SIGNAL	VARIABLE
Assignment	<=	:=
Utility	Represents circuit interconnects (wires)	Represents local information
Scope	Can be global (seen by entire code)	Local (visible only inside the corresponding PROCESS, FUNCTION, or PROCEDURE)
Behavior	Update is not immediate in sequential code (new value generally only available at the conclusion of the PROCESS, FUNCTION, or PROCEDURE)	Updated immediately (new value can be used in the next line of code)
Usage	In a PACKAGE, ENTITY, or ARCHITECTURE. In an ENTITY, all PORTS are SIGNALS by default	Only in sequential code, that is, in a PROCESS, FUNCTION, or PROCEDURE

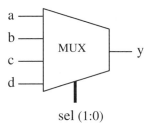

Figure 7.2
Multiplexer of example 7.3.

We want to stress again that an assignment to a VARIABLE is immediate, but that is not the case with a SIGNAL. In general, the new value of a SIGNAL will only be available at the conclusion of the current run of the corresponding PROCESS. Though this might not be always the case, it is a safe practice to consider it so. The examples presented below will further illustrate this and other differences between SIGNALS and VARIABLES.

Example 7.3:   Bad versus Good Multiplexer

In this example, we will implement the same multiplexer of example 5.2 (repeated in figure 7.2). This is, indeed, a classical example regarding the choice of a SIGNAL versus a VARIABLE.

```
1  -- Solution 1: using a SIGNAL (not ok) --
2  LIBRARY ieee;
3  USE ieee.std_logic_1164.all;
4  ----------------------------------------
5  ENTITY mux IS
6     PORT ( a, b, c, d, s0, s1: IN STD_LOGIC;
7             y: OUT STD_LOGIC);
8  END mux;
9  ----------------------------------------
10 ARCHITECTURE not_ok OF mux IS
11    SIGNAL sel : INTEGER RANGE 0 TO 3;
12 BEGIN
13    PROCESS (a, b, c, d, s0, s1)
14    BEGIN
15       sel <= 0;
16       IF (s0='1') THEN sel <= sel + 1;
```

```
17        END IF;
18        IF (s1='1') THEN sel <= sel + 2;
19        END IF;
20        CASE sel IS
21            WHEN 0 => y<=a;
22            WHEN 1 => y<=b;
23            WHEN 2 => y<=c;
24            WHEN 3 => y<=d;
25        END CASE;
26    END PROCESS;
27 END not_ok;
28 -----------------------------------------

1  -- Solution 2: using a VARIABLE (ok) ----
2  LIBRARY ieee;
3  USE ieee.std_logic_1164.all;
4  -----------------------------------------
5  ENTITY mux IS
6     PORT ( a, b, c, d, s0, s1: IN STD_LOGIC;
7             y: OUT STD_LOGIC);
8  END mux;
9  -----------------------------------------
10 ARCHITECTURE ok OF mux IS
11 BEGIN
12    PROCESS (a, b, c, d, s0, s1)
13       VARIABLE sel : INTEGER RANGE 0 TO 3;
14    BEGIN
15       sel := 0;
16       IF (s0='1') THEN sel := sel + 1;
17       END IF;
18       IF (s1='1') THEN sel := sel + 2;
19       END IF;
20       CASE sel IS
21           WHEN 0 => y<=a;
22           WHEN 1 => y<=b;
23           WHEN 2 => y<=c;
24           WHEN 3 => y<=d;
25       END CASE;
```

```
26      END PROCESS;
27 END ok;
28 -------------------------------------
```

Comments:

A common mistake when using a SIGNAL is not to remember that it might require a certain amount of time to be updated. Therefore, the assignment sel $<=$ sel $+ 1$ in the first solution (line 16) will result in one plus whatever value had been previously propagated to sel, for the assignment sel $<= 0$ (line 15) might not have had time to propagate yet. The same is true for sel $<=$ sel $+ 2$ (line 18). This is not a problem when using a VARIABLE, for its assignment is always immediate.

A second aspect that might be a problem in solution 1 is that more than one assignment is being made to the same SIGNAL (sel, lines 15, 16, and 18), which might not be acceptable. Generally, only one assignment to a SIGNAL is allowed within a PROCESS, so the software will either consider only the last one (sel $<=$ sel $+ 2$ in solution 1) or simply issue an error message and stop compilation. Again, this is never a problem when using a VARIABLE.

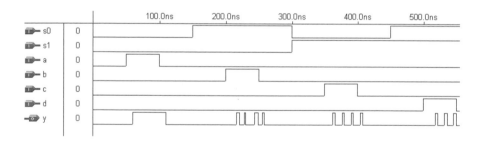

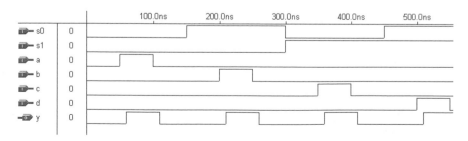

Figure 7.3
Simulation results of example 7.3.

Simulation results from both solutions are shown in figure 7.3 (bad mux in the upper graph, good mux in the lower graph). As can be seen, only solution 2 works properly.

Example 7.4:   DFF with q and qbar #1

We want to implement the DFF of figure 7.4. This circuit differs from that of example 6.1 by the absence of reset and the inclusion of qbar. The presence of qbar will help understand how an assignment to a SIGNAL is made (recall that a PORT is a SIGNAL by default).

```
1   ---- Solution 1: not OK ---------------
2   LIBRARY ieee;
3   USE ieee.std_logic_1164.all;
4   ----------------------------------------
5   ENTITY dff IS
6      PORT ( d, clk: IN STD_LOGIC;
7              q: BUFFER STD_LOGIC;
8              qbar: OUT STD_LOGIC);
9   END dff;
10  ----------------------------------------
11  ARCHITECTURE not_ok OF dff IS
12  BEGIN
13     PROCESS (clk)
14     BEGIN
15        IF (clk'EVENT AND clk='1') THEN
16           q <= d;
17           qbar <= NOT q;
18        END IF;
19     END PROCESS;
20  END not_ok;
21  ----------------------------------------
```

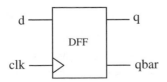

Figure 7.4
DFF of example 7.4.

```
1    ---- Solution 2: OK -------------------
2    LIBRARY ieee;
3    USE ieee.std_logic_1164.all;
4    ---------------------------------------
5    ENTITY dff IS
6       PORT ( d, clk: IN STD_LOGIC;
7                q: BUFFER STD_LOGIC;
8                qbar: OUT STD_LOGIC);
9    END dff;
10   ---------------------------------------
11   ARCHITECTURE ok OF dff IS
12   BEGIN
13      PROCESS (clk)
14      BEGIN
15         IF (clk'EVENT AND clk='1') THEN
16            q <= d;
17         END IF;
18      END PROCESS;
19      qbar <= NOT q;
20   END ok;
21   ---------------------------------------
```

Comments:

In solution 1, the assignments q<=d (line 16) and qbar<=NOT q (line 17) are both synchronous, so their new values will only be available at the conclusion of the PROCESS. This is a problem for qbar, because the new value of q has not propagated yet. Therefore, qbar will assume the reverse of the old value of q. In other words, the right value of qbar will be one clock cycle delayed, thus causing the circuit not to work correctly. This behavior can be observed in the upper graph of figure 7.5.

In solution 2, we have placed qbar<=NOT q (line 30) outside the PROCESS, thus operating as a true concurrent expression. The behavior of the resulting circuit can be observed in the lower graph of figure 7.5.

Example 7.5:   Frequency Divider

In this example, we want to implement a circuit that divides the clock frequency by 6 (figure 7.6). Intentionally, we have implemented two outputs, one based on a SIGNAL (count1) and the other based on a VARIABLE (count2). Knowing that both work properly (see simulation results in figure 7.7), you are invited to fill in the two blanks and to explain your answers.

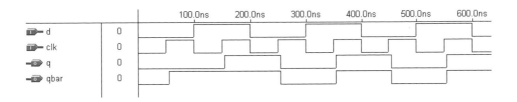

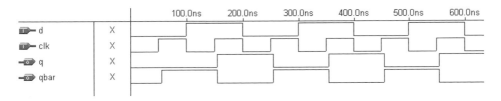

Figure 7.5
Simulation results of example 7.4.

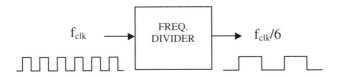

Figure 7.6
Frequency divider of example 7.5.

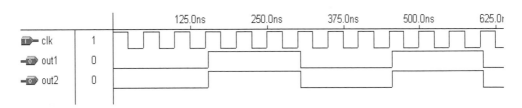

Figure 7.7
Simulation results of example 7.5.

```
1  ----------------------------------------
2  LIBRARY ieee;
3  USE ieee.std_logic_1164.all;
4  ----------------------------------------
5  ENTITY freq_divider IS
6     PORT ( clk : IN STD_LOGIC;
7              out1, out2 : BUFFER STD_LOGIC);
8  END freq_divider;
9  ----------------------------------------
10 ARCHITECTURE example OF freq_divider IS
11    SIGNAL count1 : INTEGER RANGE 0 TO 7;
12 BEGIN
13    PROCESS (clk)
14       VARIABLE count2 : INTEGER RANGE 0 TO 7;
15    BEGIN
16       IF (clk'EVENT AND clk='1') THEN
17          count1 <= count1 + 1;
18          count2 := count2 + 1;
19          IF (count1 = ? ) THEN
20             out1 <= NOT out1;
21             count1 <= 0;
22          END IF;
23          IF (count2 = ? ) THEN
24             out2 <= NOT out2;
25             count2 := 0;
26          END IF;
27       END IF;
28    END PROCESS;
29 END example;
30 ----------------------------------------
```

7.5 Number of Registers

In this section, we will discuss the number of flip-flops inferred from the code
by the compiler. The purpose is not only to understand which approaches require
less registers, but also to make sure that the code does implement the expected
circuit.

A SIGNAL generates a flip-flop whenever an assignment is made at the transition of another signal; that is, when a synchronous assignment occurs. Such assignment, being synchronous, can only happen inside a PROCESS, FUNCTION, or PROCEDURE (usually following a declaration of the type "IF signal'EVENT ..." or "WAIT UNTIL ...").

A VARIABLE, on the other hand, will not necessarily generate flip-flops if its value never leaves the PROCESS (or FUNCTION, or PROCEDURE). However, if a value is assigned to a variable at the transition of another signal, and such value is eventually passed to a signal (which leaves the process), then flip-flops will be inferred. A VARIABLE also generates a register when it is used before a value has been assigned to it. The examples presented below will illustrate these points.

Example: In the process shown below, output1 and output2 will both be stored (that is, infer flip-flops), because both are assigned at the transition of another signal (clk).

```
PROCESS (clk)
BEGIN
    IF (clk'EVENT AND clk='1') THEN
        output1 <= temp;     -- output1 stored
        output2 <= a;        -- output2 stored
    END IF;
END PROCESS;
```

Example: In the next process, only output1 will be stored (output2 will make use of logic gates).

```
PROCESS (clk)
BEGIN
    IF (clk'EVENT AND clk='1') THEN
        output1 <= temp;     -- output1 stored
    END IF;
    output2 <= a;            -- output2 not stored
END PROCESS;
```

Example: In the process below, temp (a variable) will cause x (a signal) to be stored.

```
PROCESS (clk)
    VARIABLE temp: BIT;
BEGIN
    IF (clk'EVENT AND clk='1') THEN
```

```
      temp <= a;
   END IF;
   x <= temp;     -- temp causes x to be stored
END PROCESS;
```

Additional (complete) examples are presented next. The purpose is to further illustrate when and why registers are inferred from SIGNAL and VARIABLE assignments.

Example 7.6:   DFF with q and qbar #2

Let us consider the DFF of figure 7.4 once again. Both solutions presented below function properly. The difference between them, however, resides in the number of flip-flops needed in each case. Solution 1 has two synchronous SIGNAL assignments (lines 16–17), so 2 flip-flops will be generated. This is not the case in solution 2, where one of the assignments (line 19) is no longer synchronous. The resulting circuits are presented in figures 7.8(a)–(b), respectively.

```
1   ---- Solution 1: Two DFFs ---------------
2   LIBRARY ieee;
3   USE ieee.std_logic_1164.all;
4   ----------------------------------------
5   ENTITY dff IS
6      PORT ( d, clk: IN STD_LOGIC;
```

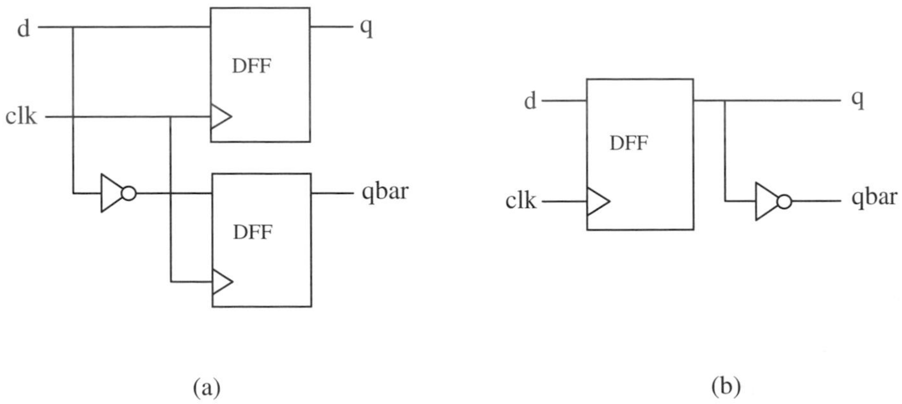

(a)   (b)

Figure 7.8
Circuits inferred from the code of example 7.6: (a) solution 1, (b) solution 2.

```
7            q: BUFFER STD_LOGIC;
8            qbar: OUT STD_LOGIC);
9  END dff;
10 -----------------------------------------
11 ARCHITECTURE two_dff OF dff IS
12 BEGIN
13    PROCESS (clk)
14    BEGIN
15      IF (clk'EVENT AND clk='1') THEN
16        q <= d;            -- generates a register
17        qbar <= NOT d;     -- generates a register
18      END IF;
19    END PROCESS;
20 END two_dff;
21 -----------------------------------------

1  ---- Solution 2: One DFF ----------------
2  LIBRARY ieee;
3  USE ieee.std_logic_1164.all;
4  -----------------------------------------
5  ENTITY dff IS
6     PORT ( d, clk: IN STD_LOGIC;
7            q: BUFFER STD_LOGIC;
8            qbar: OUT STD_LOGIC);
9  END dff;
10 -----------------------------------------
11 ARCHITECTURE one_dff OF dff IS
12 BEGIN
13    PROCESS (clk)
14    BEGIN
15      IF (clk'EVENT AND clk='1') THEN
16        q <= d;       -- generates a register
17      END IF;
18    END PROCESS;
19    qbar <= NOT q;    -- uses logic gate (no register)
20 END one_dff;
21 -----------------------------------------
```

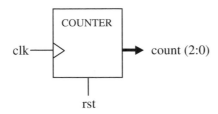

Figure 7.9
0-to-7 counter of example 7.7.

Comments:

Example 7.6 illustrates a very important situation, in which extra (unnecessary) hardware might be inferred when the code is not assembled carefully. With solution 2, the synthesizer will always infer only one flip-flop. It is interesting to mention, however, that for certain types of CPLD/FPGA devices, when the signals q and qbar are connected directly to chip pins, the fitter (place & route) might still opt for two flip-flops in the physical implementation. This does not mean that two flip-flops were indeed necessary. In fact, though the fitter (place & route) report might mention two registers in such cases, the synthesis report will invariably inform that only one register was indeed required. A further discussion is presented in problem 7.7.

Example 7.7:   Counter

Let us consider the 0-to-7 counter of figure 7.9. Two solutions are presented below. In the first, a synchronous VARIABLE assignment is made (lines 14–15). In the second, a synchronous SIGNAL assignment occurs (lines 13–14).

From either solution, three flip-flops are inferred (to hold the 3-bit output signal count). Solution 1 is an example that a VARIABLE can indeed generate registers. The reason is that its assignment (line 15) is at the transition of another signal (clk, line 14) and its value does leave the PROCESS (line 17).

Solution 2, on the other hand, uses only SIGNALS. Notice that, since no auxiliary signal was used, count needed to be declared as of mode BUFFER (line 4), because it is assigned a value *and* is also read (used) internally (line 14). Still regarding line 14 of solution 2, notice that a SIGNAL, like a VARIABLE, can also be incremented when used in a sequential code. Finally, notice that neither in solution 1 nor in solution 2 was the std_logic_1164 package declared, because we are not using std_logic data types in this example.

```
1   ------ Solution 1: With a VARIABLE --------
2   ENTITY counter IS
```

```
3     PORT ( clk, rst: IN BIT;
4            count: OUT INTEGER RANGE 0 TO 7);
5   END counter;
6   ---------------------------------------------
7   ARCHITECTURE counter OF counter IS
8   BEGIN
9      PROCESS (clk, rst)
10        VARIABLE temp: INTEGER RANGE 0 TO 7;
11     BEGIN
12        IF (rst='1') THEN
13           temp:=0;
14        ELSIF (clk'EVENT AND clk='1') THEN
15           temp := temp+1;
16        END IF;
17        count <= temp;
18     END PROCESS;
19  END counter;
20  ---------------------------------------------

1   ------ Solution 2: With SIGNALS only -------
2   ENTITY counter IS
3      PORT ( clk, rst: IN BIT;
4            count: BUFFER INTEGER RANGE 0 TO 7);
5   END counter;
6   ---------------------------------------------
7   ARCHITECTURE counter OF counter IS
8   BEGIN
9      PROCESS (clk, rst)
10     BEGIN
11        IF (rst='1') THEN
12           count <= 0;
13        ELSIF (clk'EVENT AND clk='1') THEN
14           count <= count + 1;
15        END IF;
16     END PROCESS;
17  END counter;
18  ---------------------------------------------
```

Simulation results (from either solution above) are shown in figure 7.10.

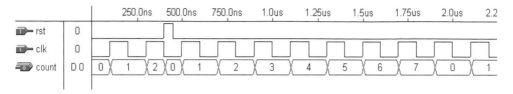

Figure 7.10
Simulation results of example 7.7.

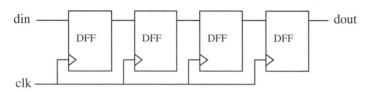

Figure 7.11
Shift-register of example 7.8.

Example 7.8:   Shift Register #1

We are now interested in examining what happens to the 4-stage shift register of figure 7.11 when different VARIABLE and SIGNAL assignments are made. Of course, if the solution is correct, then the output signal (dout) should be four positive clock edges behind the input signal (din).

In solution 1, three VARIABLES are used (a, b, and c, line 10). However, the variables are used before values are assigned to them (that is, in reverse order, starting with dout, line 13, and ending with din, line 16). Consequently, flip-flops will be inferred, which store the values from the previous run of the PROCESS.

In solution 2, the variables were replaced by SIGNALS (line 8), and the assignments are made in direct order (from din to dout, lines 13–16). Since signal assignments at the transition of another signal do generate registers, here too the right circuit will be inferred.

Finally, in solution 3, the same variables of solution 1 were employed, but in direct order (from din to dout, lines 13–16). Recall, however, that an assignment to a variable is immediate, and since the variables are being used in direct order (that is, after values have been assigned to them), lines 13–15 collapse into one line, equivalent to c := din. The value of c does leave the process in the next line (line 16), however, where a signal assignment (dout <= c) occurs at the transition of clk. Therefore, one register will be inferred from solution 3, thus not resulting the correct circuit.

Note: More conventional solutions to the shift-register problem will be presented in example 7.9.

```
1   -------- Solution 1: -----------------
2   ENTITY shift IS
3      PORT ( din, clk: IN BIT;
4              dout: OUT BIT);
5   END shift;
6   ------------------------------------
7   ARCHITECTURE shift OF shift IS
8   BEGIN
9      PROCESS (clk)
10        VARIABLE a, b, c: BIT;
11     BEGIN
12        IF (clk'EVENT AND clk='1') THEN
13           dout <= c;
14           c := b;
15           b := a;
16           a := din;
17        END IF;
18     END PROCESS;
19  END shift;
20  ------------------------------------

1   -------- Solution 2: ----------------
2   ENTITY shift IS
3      PORT ( din, clk: IN BIT;
4              dout: OUT BIT);
5   END shift;
6   ------------------------------------
7   ARCHITECTURE shift OF shift IS
8      SIGNAL a, b, c: BIT;
9   BEGIN
10     PROCESS (clk)
11     BEGIN
12        IF (clk'EVENT AND clk='1') THEN
13           a <= din;
14           b <= a;
```

```
15          c <= b;
16          dout <= c;
17       END IF;
18    END PROCESS;
19 END shift;
20 -------------------------------------

1  -------- Solution 3: ----------------
2  ENTITY shift IS
3     PORT ( din, clk: IN BIT;
4             dout: OUT BIT);
5  END shift;
6  -------------------------------------
7  ARCHITECTURE shift OF shift IS
8  BEGIN
9     PROCESS (clk)
10       VARIABLE a, b, c: BIT;
11    BEGIN
12       IF (clk'EVENT AND clk='1') THEN
13          a := din;
14          b := a;
15          c := b;
16          dout <= c;
17       END IF;
18    END PROCESS;
19 END shift;
20 -------------------------------------
```

Simulation results from solution 1 or 2 are shown in the upper graph of figure 7.12, while the lower graph shows results from solution 3. As expected, dout is four positive clock edges behind din in the former, but only one positive edge behind the input in the latter.

Example 7.9: Shift Register #2

In this example, *conventional* approaches to the design of shift registers are presented.

Figure 7.13 shows a 4-bit shift register, similar to that of example 7.8, except for the presence of a reset input (rst). As before, the output bit (q) should be four positive clock edges behind the input bit (d). Reset should be asynchronous, forcing all flip-flop outputs to '0' when asserted.

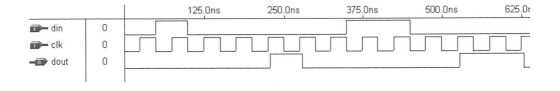

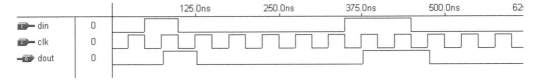

Figure 7.12
Simulation results of example 7.8 (solutions 1 and 2).

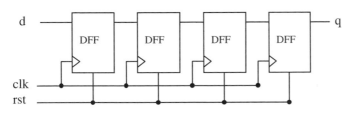

Figure 7.13
Shift register of example 7.9.

Two solutions are presented. One uses a SIGNAL to generate the flip-flops, while the other uses a VARIABLE. The synthesized circuits are the same (that is, four flip-flops are inferred from either solution). In solution 1, registers are created because an assignment to a signal is made at the transition of another signal (lines 17–18). In solution 2, the assignment at the transition of another signal is made to a variable (lines 17–18), but since its value does leave the process (that is, it is passed to a port in line 20), it too infers registers.

```
1   ---- Solution 1: With an internal SIGNAL ---
2   LIBRARY ieee;
3   USE ieee.std_logic_1164.all;
4   ------------------------------------------------
5   ENTITY shiftreg IS
```

```
6     PORT ( d, clk, rst: IN STD_LOGIC;
7           q: OUT STD_LOGIC);
8  END shiftreg;
9  -------------------------------------------
10 ARCHITECTURE behavior OF shiftreg IS
11    SIGNAL internal: STD_LOGIC_VECTOR (3 DOWNTO 0);
12 BEGIN
13    PROCESS (clk, rst)
14    BEGIN
15      IF (rst='1') THEN
16          internal <= (OTHERS => '0');
17      ELSIF (clk'EVENT AND clk='1') THEN
18          internal <= d & internal(3 DOWNTO 1);
19      END IF;
20    END PROCESS;
21    q <= internal(0);
22 END behavior;
23 -------------------------------------------

1  -- Solution 2: With an internal VARIABLE ---
2  LIBRARY ieee;
3  USE ieee.std_logic_1164.all;
4  -------------------------------------------
5  ENTITY shiftreg IS
6     PORT ( d, clk, rst: IN STD_LOGIC;
7           q: OUT STD_LOGIC);
8  END shiftreg;
9  -------------------------------------------
10 ARCHITECTURE behavior OF shiftreg IS
11 BEGIN
12    PROCESS (clk, rst)
13      VARIABLE internal: STD_LOGIC_VECTOR (3 DOWNTO 0);
14    BEGIN
15      IF (rst='1') THEN
16          internal := (OTHERS => '0');
17      ELSIF (clk'EVENT AND clk='1') THEN
18          internal := d & internal(3 DOWNTO 1);
19      END IF;
20    q <= internal(0);
```

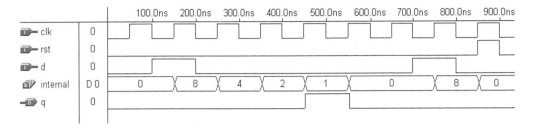

Figure 7.14
Simulation results of example 7.9.

```
21    END PROCESS;
22 END behavior;
23 -------------------------------------------
```

Simulation results (from either solution above) are shown in figure 7.14. As can be seen, q is indeed four positive clock edges behind d.

You may now review the usage of SIGNALS and VARIABLES in all examples of chapter 6. Moreover, in chapter 8, a series of design examples will be presented in which the correct understanding of the differences between signals and variables is crucial, or the wrong circuit might be inferred.

7.6   Problems

Problem 7.1:   VHDL "Numerical" Objects

Given the following VHDL objects:

```
CONSTANT max : INTEGER := 10;
SIGNAL x: INTEGER RANGE -10 TO 10;
SIGNAL y: BIT_VECTOR (15 DOWNTO 0);
VARIABLE z: BIT;
```

Determine which among the assignments below are legal (suggestion: review chapter 3).

```
x <= 5;
x <= y(5);
z <= '1';
z := y(5);
WHILE i IN 0 TO max LOOP...
```

152

Chapter 7

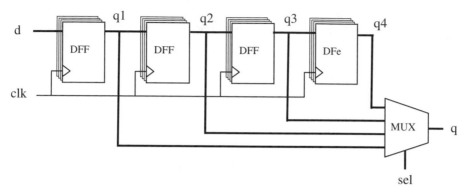

Figure P7.2

```
FOR i IN 0 TO x LOOP...
G1: FOR i IN 0 TO max GENERATE...
G1: FOR i IN 0 TO x GENERATE...
```

Problem 7.2:   Data Delay

Figure P7.2 shows the diagram of a programmable data delay circuit. The input (d) and output (q) are 4-bit buses. Depending on the value of sel (select), q should be one, two, three, or four clock cycles delayed with respect to d.

(a) Write a VHDL code for this circuit;

(b) How many flip-flops do you expect your solution to contain?

(c) Synthesize your solution and open the report file. Verify whether the actual number of flip-flops matches your prediction.

Problem 7.3:   DFF with q and qbar #1

We want to implement the same flip-flop of example 7.4 (figure 7.4). However, we have introduced an auxiliary signal (temp) in our code. You are asked to examine each of the solutions below and determine whether q and qbar will work properly. Briefly explain your answers.

```
------------------------------------
ENTITY dff IS
   PORT ( d, clk: IN BIT;
          q, qbar: BUFFER BIT);
END dff;
```

```
-------- Solution 1 -------------------
ARCHITECTURE arch1 OF dff IS
   SIGNAL temp: BIT;
BEGIN
   PROCESS (clk)
   BEGIN
      IF (clk'EVENT AND clk='1') THEN
         temp <= d;
         q <= temp;
         qbar <= NOT temp;
      END IF;
   END PROCESS;
END arch1;
-------- Solution 2 -------------------
ARCHITECTURE arch2 OF dff IS
   SIGNAL temp: BIT;
BEGIN
   PROCESS (clk)
   BEGIN
      IF (clk'EVENT AND clk='1') THEN
         temp <= d;
      END IF;
      q <= temp;
      qbar <= NOT temp;
   END PROCESS;
END arch2;
-------- Solution 3 -------------------
ARCHITECTURE arch3 OF dff IS
   SIGNAL temp: BIT;
BEGIN
   PROCESS (clk)
   BEGIN
      IF (clk'EVENT AND clk='1') THEN
         temp <= d;
      END IF;
   END PROCESS;
   q <= temp;
   qbar <= NOT temp;
```

```
END arch3;
```

Problem 7.4:   DFF with q and qbar #2

This problem is similar to problem 7.3. However, here we have an auxiliary VARI-
ABLE instead of an auxiliary SIGNAL. You are asked to examine each of the
solutions below and determine whether q and qbar will work as expected. Briefly
explain your answers.

```
----------------------------------------
ENTITY dff IS
    PORT ( d, clk: IN BIT;
           q: BUFFER BIT;
           qbar: OUT BIT);
END dff;
-------- Solution 1 --------------------
ARCHITECTURE arch1 OF dff IS
BEGIN
    PROCESS (clk)
       VARIABLE temp: BIT;
    BEGIN
       IF (clk'EVENT AND clk='1') THEN
          temp := d;
          q <= temp;
          qbar <= NOT temp;
       END IF;
    END PROCESS;
END arch1;
-------- Solution 2 --------------------
ARCHITECTURE arch2 OF dff IS
BEGIN
    PROCESS (clk)
       VARIABLE temp: BIT;
    BEGIN
       IF (clk'EVENT AND clk='1') THEN
          temp := d;
          q <= temp;
          qbar <= NOT q;
```

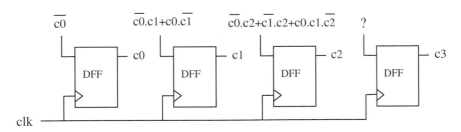

Figure P7.5

```
      END IF;
   END PROCESS;
END arch2;
-------- Solution 3 --------------------
ARCHITECTURE arch3 OF dff IS
BEGIN
   PROCESS (clk)
      VARIABLE temp: BIT;
   BEGIN
      IF (clk'EVENT AND clk='1') THEN
         temp := d;
         q <= temp;
      END IF;
   END PROCESS;
   qbar <= NOT q;
END arch3;
-------------------------------------
```

Problem 7.5:   Counter

Consider the 4-bit counter of example 6.2. However, suppose that now it should count from 0 ("0000") to 15 ("1111").

(a) Write a VHDL code for it, then synthesize and simulate your solution to verify that it works as expected.

(b) Open the report file created by your synthesis tool and confirm that four flip-flops were inferred.

(c) Still using the report file, observe whether the circuit looks like that of figure P7.5. Are the equations implemented at the flip-flop inputs similar or equivalent to those shown in figure P7.5? What is the missing equation (input of fourth flip-flop)?

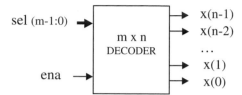

Figure P7.6

Problem 7.6:   Generic n-by-m Decoder

Let us consider the generic n-by-m decoder presented in example 4.1 (repeated in figure P7.6). The code presented below, though very compact, contains a flaw in the assignment "x<=(sel=>'0', OTHERS=>'1'". The reason is that sel is not a locally stable signal (indeed, it appears in the sensitivity list of the PROCESS). You are asked to correct the code.

```
---------------------------------------------
LIBRARY ieee;
USE ieee.std_logic_1164.all;
---------------------------------------------
ENTITY decoder IS
   PORT ( ena : IN STD_LOGIC;
          sel : IN INTEGER RANGE 0 TO 7;
          x : OUT STD_LOGIC_VECTOR (7 DOWNTO 0));
END decoder;
---------------------------------------------
ARCHITECTURE not_ok OF decoder IS
BEGIN
   PROCESS (ena, sel)
   BEGIN
      IF (ena='0') THEN
         x <= (OTHERS => '1');
      ELSE
         x <= (sel=>'0', OTHERS => '1');
      END IF;
   END PROCESS;
END not_ok;
---------------------------------------------
```

Problem 7.7: DFF with q and qbar #3

Consider the DFF implemented in solution 2 of example 7.6. We are interested in examining the number of registers required in its implementation. We already know that the answer is *one*. However, as we mentioned in the comments of example 7.6, even though the synthesizer tells us so, the fitter (place & route) might opt for two registers in the final (physical) implementation when q and qbar are connected directly to output pins. This problem deals with this kind of situation.

(a) Compile the code of example 7.6 (solution 2) using Quartus II 3.0 (appendix D). Select a device from the MAX3000A or Cyclone family. In the synthesis reports, verify the number of registers inferred and the equations implemented by the synthesizer (confirming the number of flip-flops). Next, repeat these verifications in the fitter reports (number of registers and equations).

(b) Repeat the procedure above for another device. Select a chip from the FLEX10K family.

(c) Compile now the code of example 7.6 (solution 2) using ISE 6.1 (appendix B). Select a device from the XC9500 or CoolRunner II family. After compilation, make the same verifications described above.

(d) Finally, consider the case when one of the outputs of the flip-flop is not connected directly to a pin. In order to do so, we have introduced a signal called *test* in the code below. Repeat all topics above for this new code.

```
1   ------------------------------------------
2   LIBRARY ieee;
3   USE ieee.std_logic_1164.all;
4   ------------------------------------------
5   ENTITY dff IS
6      PORT ( d, clk, test: IN STD_LOGIC;
7             q: BUFFER STD_LOGIC;
8             qbar: OUT STD_LOGIC);
9   END dff;
10  ------------------------------------------
11  ARCHITECTURE one_dff OF dff IS
12  BEGIN
13     PROCESS (clk)
14     BEGIN
```

```
15      IF (clk'EVENT AND clk='1') THEN
16         q <= d;
17      END IF;
18   END PROCESS;
19   qbar <= NOT q AND test;
20 END one_dff;
21 ----------------------------------------
```

8 State Machines

Finite state machines (FSM) constitute a special modeling technique for sequential logic circuits. Such a model can be very helpful in the design of certain types of systems, particularly those whose tasks form a well-defined sequence (digital controllers, for example). We start the chapter by reviewing fundamental concepts related to FSM. We then introduce corresponding VHDL coding techniques, followed by complete design examples.

8.1 Introduction

Figure 8.1 shows the block diagram of a single-phase state machine. As indicated in the figure, the lower section contains the sequential logic (flip-flops), while the upper section contains the combinational logic.

The combinational (upper) section has two inputs, being one pr_state (present state) and the other the external input proper. It has also two outputs, nx_state (next state) and the external output proper.

The sequential (lower) section has three inputs (clock, reset, and nx_state), and one output (pr_state). Since all flip-flops are in this part of the system, clock and reset must be connected to it.

If the output of the machine depends not only on the present state but also on the current input, then it is called a Mealy machine. Otherwise, if it depends only on the current state, it is called a Moore machine. Examples of both will be shown later.

The separation of the circuit into two sections (figure 8.1) allows the design to be broken into two parts as well. From a VHDL perspective, it is clear that the lower part, being sequential, will require a PROCESS, while the upper part, being combinational, will not. However, recall that sequential code can implement both types of logic, combinational as well as sequential. Hence, if desired, the upper part can also be implemented using a PROCESS.

The signals clock and reset normally appear in the sensitivity list of the lower section's PROCESS (unless reset is synchronous or not used, or WAIT is used instead of IF). When reset is asserted, pr_state will be set to the system's initial state. Otherwise, at the proper clock edge the flip-flops will store nx_state, thus transferring it to the lower section's output (pr_state).

One important aspect related to the FSM approach is that, though any sequential circuit can in principle be modeled as a state machine, this is not always advantageous. The reason is that the code might become longer, more complex, and more error prone than in a conventional approach. This is often the case with simple registered

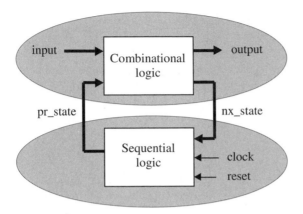

Figure 8.1
Mealy (Moore) state machine diagram.

circuits, like counters. As a simple rule of thumb, the FSM approach is advisable in systems whose tasks constitute a well-structured list so all states can be easily enumerated. That is, in a typical state machine implementation, we will encounter, at the beginning of the ARCHITECTURE, a user-defined *enumerated* data type, containing a list of all possible system states. Digital controllers are good examples of such circuits.

Another important aspect, which was already emphasized at the beginning of chapter 5, is that not all circuits that possess memory are necessarily sequential. A RAM (Random Access Memory) was given as an example. In it, the memory-read operation depends only on the address bits presently applied to the RAM (current input), with the retrieved value having nothing to do with previous memory accesses (previous inputs). In such cases, the FSM approach is not advisable.

8.2   Design Style #1

Several approaches can be conceived to design a FSM. We will describe in detail one style that is well structured and easily applicable. In it, the design of the lower section of the state machine (figure 8.1) is completely separated from that of the upper section. All states of the machine are always explicitly declared using an enumerated data type. After introducing such a design style, we will examine it from a data storage perspective, in order to further understand and refine its construction, which will lead to design style #2.

Design of the Lower (Sequential) Section

In figure 8.1, the flip-flops are in the lower section, so clock and reset are connected to it. The other lower section's input is nx_state (next state), while pr_state (present state) is its only output. Being the circuit of the lower section sequential, a PROCESS is required, in which any of the sequential statements (IF, WAIT, CASE, or LOOP, chapter 6) can be employed.

A typical design template for the lower section is the following:

```
PROCESS (reset, clock)
BEGIN
   IF (reset='1') THEN
      pr_state <= state0;
   ELSIF (clock'EVENT AND clock='1') THEN
      pr_state <= nx_state;
   END IF;
END PROCESS;
```

The code shown above is very simple. It consists of an asynchronous reset, which determines the initial state of the system (state0), followed by the synchronous storage of nx_state (at the positive transition of clock), which will produce pr_state at the lower section's output (figure 8.1). One good thing about this approach is that the design of the lower section is basically standard.

Another advantage of this design style is that the number of registers is minimum. From section 7.5, we know that the number of flip-flops inferred from the code above is simply equal to the number of bits needed to encode all states of the FSM (because the only signal to which a value is assigned at the transition of another signal is pr_state). Therefore, if the default (binary) encoding style (section 8.4) is used, just $\lceil \log_2 n \rceil$ flip-flops will then be needed, where n is the number of states.

Design of the Upper (Combinational) Section

In figure 8.1, the upper section is fully combinational, so its code does not need to be sequential; concurrent code can be used as well. Yet, in the design template shown below, sequential code was employed, with the CASE statement playing the central role. In this case, recall that rules 1 and 2 of section 6.10 must be observed.

```
PROCESS (input, pr_state)
BEGIN
   CASE pr_state IS
      WHEN state0 =>
         IF (input = ...) THEN
            output <= <value>;
            nx_state <= state1;
         ELSE ...
         END IF;
      WHEN state1 =>
         IF (input = ...) THEN
            output <= <value>;
            nx_state <= state2;
         ELSE ...
         END IF;
      WHEN state2 =>
         IF (input = ...) THEN
            output <= <value>;
            nx_state <= state2;
         ELSE ...
         END IF;
      ...
   END CASE;
END PROCESS;
```

As can be seen, this code is also very simple, and does two things: (a) it assigns the output value and (b) it establishes the next state. Notice also that it complies with rules 1 and 2 of section 6.10, relative to the design of combinational circuits using sequential statements, for all input signals are present in the sensitivity list and all input/output combinations are specified. Finally, observe that no signal assignment is made at the transition of another signal, so no flip-flops will be inferred (section 7.5).

State Machine Template for Design Style #1

A complete template is shown below. Notice that, in addition to the two processes presented above, it also contains a user-defined enumerated data type (here called *state*), which lists all possible states of the machine.

```
LIBRARY ieee;
USE ieee.std_logic_1164.all;
-------------------------------------------------------
ENTITY <entity_name> IS
    PORT ( input: IN <data_type>;
           reset, clock: IN STD_LOGIC;
           output: OUT <data_type>);
END <entity_name>;
-------------------------------------------------------
ARCHITECTURE <arch_name> OF <entity_name> IS
    TYPE state IS (state0, state1, state2, state3, ...);
    SIGNAL pr_state, nx_state: state;
BEGIN
    ---------- Lower section: ----------------------
    PROCESS (reset, clock)
    BEGIN
        IF (reset='1') THEN
           pr_state <= state0;
        ELSIF (clock'EVENT AND clock='1') THEN
           pr_state <= nx_state;
        END IF;
    END PROCESS;
    ---------- Upper section: ----------------------
    PROCESS (input, pr_state)
    BEGIN
        CASE pr_state IS
            WHEN state0 =>
                IF (input = ...) THEN
                    output <= <value>;
                    nx_state <= state1;
                ELSE ...
                END IF;
            WHEN state1 =>
                IF (input = ...) THEN
                    output <= <value>;
                    nx_state <= state2;
                ELSE ...
                END IF;
            WHEN state2 =>
                IF (input = ...) THEN
                    output <= <value>;
                    nx_state <= state3;
                ELSE ...
                END IF;
            ...
        END CASE;
    END PROCESS;
END <arch_name>;
```

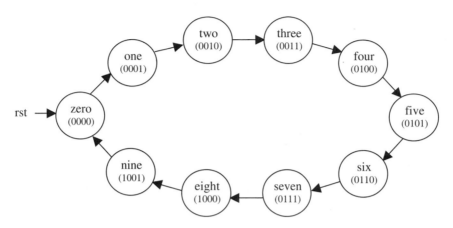

Figure 8.2
States diagram of example 8.1.

Example 8.1:   BCD Counter

A counter is an example of Moore machine, for the output depends only on the stored (present) state. As a simple registered circuit and as a sequencer, it can be easily implemented in either approach: conventional (as we have already done in previous chapters) or FSM type. The problem with the latter is that when the number of states is large it becomes cumbersome to enumerate them all, a problem easily avoided using the LOOP statement in a conventional approach.

The state diagram of a 0-to-9 circular counter is shown in figure 8.2. The states were called zero, one, ..., nine, each name corresponding to the decimal value of the output.

A VHDL code, directly resembling the design style #1 template, is presented below. An enumerated data type (state) appears in lines 11–12. The design of the lower (clocked) section is presented in lines 16–23, and that of the upper (combinational) section, in lines 25–59. In this example, the number of registers is $\lceil \log_2 10 \rceil = 4$.

Simulation results are shown in figure 8.3. As can be seen, the output (count) grows from 0 to 9, and then restarts from 0 again.

```
1    ---------------------------------------------
2    LIBRARY ieee;
3    USE ieee.std_logic_1164.all;
4    ---------------------------------------------
5    ENTITY counter IS
```

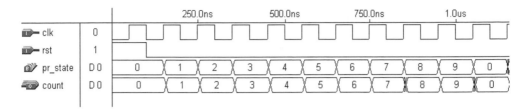

Figure 8.3
Simulation results of example 8.1.

```
6      PORT ( clk, rst: IN STD_LOGIC;
7             count: OUT STD_LOGIC_VECTOR (3 DOWNTO 0));
8   END counter;
9   ---------------------------------------------------
10  ARCHITECTURE state_machine OF counter IS
11    TYPE state IS (zero, one, two, three, four,
12        five, six, seven, eight, nine);
13    SIGNAL pr_state, nx_state: state;
14  BEGIN
15    ------------- Lower section: -----------------
16    PROCESS (rst, clk)
17    BEGIN
18      IF (rst='1') THEN
19         pr_state <= zero;
20      ELSIF (clk'EVENT AND clk='1') THEN
21         pr_state <= nx_state;
22      END IF;
23    END PROCESS;
24    ------------- Upper section: -----------------
25    PROCESS (pr_state)
26    BEGIN
27      CASE pr_state IS
28        WHEN zero =>
29           count <= "0000";
30           nx_state <= one;
31        WHEN one =>
32           count <= "0001";
33           nx_state <= two;
```

```
34          WHEN two =>
35             count <= "0010";
36             nx_state <= three;
37          WHEN three =>
38             count <= "0011";
39             nx_state <= four;
40          WHEN four =>
41             count <= "0100";
42             nx_state <= five;
43          WHEN five =>
44             count <= "0101";
45             nx_state <= six;
46          WHEN six =>
47             count <= "0110";
48             nx_state <= seven;
49          WHEN seven =>
50             count <= "0111";
51             nx_state <= eight;
52          WHEN eight =>
53             count <= "1000";
54             nx_state <= nine;
55          WHEN nine =>
56             count <= "1001";
57             nx_state <= zero;
58       END CASE;
59    END PROCESS;
60 END state_machine;
61 ------------------------------------------------
```

Example 8.2:   Simple FSM #1

Figure 8.4 shows the states diagram of a very simple FSM. The system has two states (stateA and stateB), and must change from one to the other every time d = '1' is received. The desired output is x = a when the machine is in stateA, or x = b when in stateB. The initial (reset) state is stateA.

A VHDL code for this circuit, employing design style #1, is shown below.

```
1  ------------------------------------------------
2  ENTITY simple_fsm IS
```

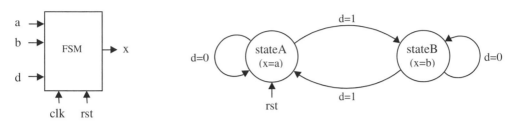

Figure 8.4
State machine of example 8.1.

```
3      PORT ( a, b, d, clk, rst: IN BIT;
4              x: OUT BIT);
5   END simple_fsm;
6   ------------------------------------------------
7   ARCHITECTURE simple_fsm OF simple_fsm IS
8      TYPE state IS (stateA, stateB);
9      SIGNAL pr_state, nx_state: state;
10  BEGIN
11     ----- Lower section: ----------------------
12     PROCESS (rst, clk)
13     BEGIN
14       IF (rst='1') THEN
15          pr_state <= stateA;
16       ELSIF (clk'EVENT AND clk='1') THEN
17          pr_state <= nx_state;
18       END IF;
19     END PROCESS;
20     ---------- Upper section: -----------------
21     PROCESS (a, b, d, pr_state)
22     BEGIN
23       CASE pr_state IS
24          WHEN stateA =>
25             x <= a;
26             IF (d='1') THEN nx_state <= stateB;
27             ELSE nx_state <= stateA;
28             END IF;
29          WHEN stateB =>
30             x <= b;
```

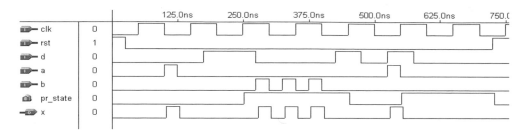

Figure 8.5
Simulation results of example 8.2

```
31                    IF (d='1') THEN nx_state <= stateA;
32                    ELSE nx_state <= stateB;
33                    END IF;
34          END CASE;
35     END PROCESS;
36 END simple_fsm;
37 ------------------------------------------------
```

Simulation results relative to the code above are shown in figure 8.5. Notice that the circuit works as expected. Indeed, looking at the report files, one will verify that, as expected, only one flip-flop was required to implement this circuit because there are only two states to be encoded. Notice also that the upper section is indeed combinational, for the output (x), which in this case does depend on the inputs (a or b, depending on which state the machine is in), varies when a or b vary, regardless of clk. If a synchronous output were required, then design style #2 should be employed.

8.3   Design Style #2 (Stored Output)

As we have seen, in design style #1 only pr_state is stored. Therefore, the overall circuit can be summarized as in figure 8.6(a). Notice that in this case, if it is a Mealy machine (one whose output is dependent on the current input), the output might change when the input changes (asynchronous output).

In many applications, the signals are required to be synchronous, so the output should be updated only when the proper clock edge occurs. To make Mealy machines synchronous, the output must be stored as well, as shown in figure 8.6(b). This structure is the object of design style #2.

To implement this new structure, very few modifications are needed. For example, we can use an additional signal (say, temp) to compute the output value (upper sec-

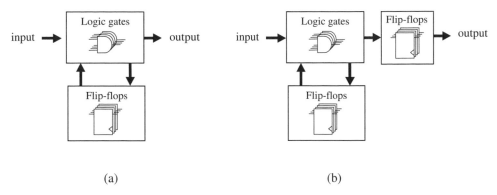

Figure 8.6
Circuit diagrams for (a) Design Style #1 and (b) Design Style #2.

tion), but only pass its value to the actual output signal when a clock event occurs (lower section). These modifications can be observed in the template shown below.

State Machine Template for Design Style #2

```
LIBRARY ieee;
USE ieee.std_logic_1164.all;
----------------------------------------------------------
ENTITY <ent_name> IS
   PORT (input: IN <data_type>;
         reset, clock: IN STD_LOGIC;
         output: OUT <data_type>);
END <ent_name>;
----------------------------------------------------------
ARCHITECTURE <arch_name> OF <ent_name> IS
   TYPE states IS (state0, state1, state2, state3, ...);
   SIGNAL pr_state, nx_state: states;
   SIGNAL temp: <data_type>;
BEGIN
   ---------- Lower section: -------------------------
   PROCESS (reset, clock)
   BEGIN
      IF (reset='1') THEN
         pr_state <= state0;
      ELSIF (clock'EVENT AND clock='1') THEN
         output <= temp;
         pr_state <= nx_state;
      END IF;
   END PROCESS;
```

```
    ---------- Upper section: --------------------------
   PROCESS (pr_state)
   BEGIN
      CASE pr_state IS
         WHEN state0 =>
            temp <= <value>;
            IF (condition) THEN nx_state <= state1;
            ...
            END IF;
         WHEN state1 =>
            temp <= <value>;
            IF (condition) THEN nx_state <= state2;
            ...
            END IF;
         WHEN state2 =>
            temp <= <value>;
            IF (condition) THEN nx_state <= state3;
            ...
            END IF;
         ...
      END CASE;
   END PROCESS;
 END <arch_name>;
```

Comparing the template of design style #2 with that of design style #1, we verify that the only differences are those related to the introduction of the internal signal temp. This signal will cause the output of the state machine to be stored, for its value is passed to the output only when clk'EVENT occurs.

Example 8.3:   Simple FSM #2

Let us consider the design of example 8.2 once again. However, let us say that now we want the output to be synchronous (to change only when clock rises). Since this is a Mealy machine, design style #2 is required.

```
1  ----------------------------------------------
2  ENTITY simple_fsm IS
3     PORT ( a, b, d, clk, rst: IN BIT;
4             x: OUT BIT);
5  END simple_fsm;
6  ----------------------------------------------
7  ARCHITECTURE simple_fsm OF simple_fsm IS
8     TYPE state IS (stateA, stateB);
9     SIGNAL pr_state, nx_state: state;
10    SIGNAL temp: BIT;
```

```
11 BEGIN
12      ----- Lower section: ----------------------
13      PROCESS (rst, clk)
14      BEGIN
15         IF (rst='1') THEN
16            pr_state <= stateA;
17         ELSIF (clk'EVENT AND clk='1') THEN
18            x <= temp;
19            pr_state <= nx_state;
20         END IF;
21      END PROCESS;
22      ---------- Upper section: ------------------
23      PROCESS (a, b, d, pr_state)
24      BEGIN
25         CASE pr_state IS
26            WHEN stateA =>
27               temp <= a;
28               IF (d='1') THEN nx_state <= stateB;
29               ELSE nx_state <= stateA;
30               END IF;
31            WHEN stateB =>
32               temp <= b;
33               IF (d='1') THEN nx_state <= stateA;
34               ELSE nx_state <= stateB;
35               END IF;
36         END CASE;
37      END PROCESS;
38 END simple_fsm;
39 -------------------------------------------------
```

Looking at the report files produced by the compiler, we observe that two flip-flops were now inferred, one to encode the states of the machine, and the other to store the output.

Simulation results are shown in figure 8.7. Recall that when a signal is stored, its value will necessarily remain static between two consecutive clock edges. Therefore, if the input (a or b in the example above) changes during this interval, the change might not be observed by the circuit; moreover, when observed, it will be delayed with respect to the input (which is proper of synchronous circuits).

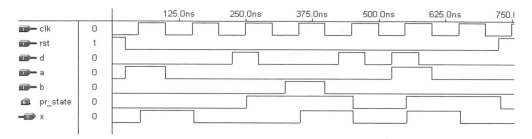

Figure 8.7
Simulation results of example 8.3.

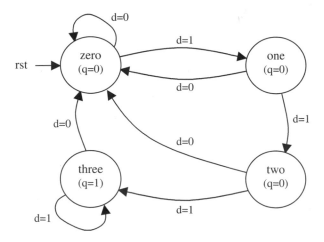

Figure 8.8
States diagram for example 8.4.

Example 8.4:   String Detector

We want to design a circuit that takes as input a serial bit stream and outputs a '1' whenever the sequence "111" occurs. Overlaps must also be considered, that is, if . . . 0111110 . . . occurs, than the output should remain active for three consecutive clock cycles.

The state diagram of our machine is shown in figure 8.8. There are four states, which we called zero, one, two, and three, with the name corresponding to the number of consecutive '1's detected. The solution shown below utilizes design style #1.

```
1   ---------------------------------------------
2   LIBRARY ieee;
3   USE ieee.std_logic_1164.all;
4   ---------------------------------------------
5   ENTITY string_detector IS
6      PORT ( d, clk, rst: IN BIT;
7               q: OUT BIT);
8   END string_detector;
9   ---------------------------------------------
10  ARCHITECTURE my_arch OF string_detector IS
11     TYPE state IS (zero, one, two, three);
12     SIGNAL pr_state, nx_state: state;
13  BEGIN
14     ----- Lower section: --------------------
15     PROCESS (rst, clk)
16     BEGIN
17        IF (rst='1') THEN
18           pr_state <= zero;
19        ELSIF (clk'EVENT AND clk='1') THEN
20           pr_state <= nx_state;
21        END IF;
22     END PROCESS;
23     ---------- Upper section: ---------------
24     PROCESS (d, pr_state)
25     BEGIN
26        CASE pr_state IS
27           WHEN zero =>
28              q <= '0';
29              IF (d='1') THEN nx_state <= one;
30              ELSE nx_state <= zero;
31              END IF;
32           WHEN one =>
33              q <= '0';
34              IF (d='1') THEN nx_state <= two;
35              ELSE nx_state <= zero;
36              END IF;
37           WHEN two =>
38              q <= '0';
39              IF (d='1') THEN nx_state <= three;
```

```
40              ELSE nx_state <= zero;
41              END IF;
42          WHEN three =>
43              q <= '1';
44              IF (d='0') THEN nx_state <= zero;
45              ELSE nx_state <= three;
46              END IF;
47      END CASE;
48  END PROCESS;
49 END my_arch;
50 -------------------------------------------
```

Notice that in this example the output does not depend on the current input. This fact can be observed in lines 28, 33, 38, and 43 of the code above, which show that all assignments to q are unconditional (that is, do not depend on d). Therefore, the output is automatically synchronous (a Moore machine), so the use of design style #2 is unnecessary. The circuit requires two flip-flops, which encode the four states of the state machine, from which q is computed.

Simulation results are shown in figure 8.9. As can be seen, the data sequence d = "011101100" was applied to the circuit, resulting the response q = "000100000" at the output.

Example 8.5: Traffic Light Controller (TLC)

As mentioned earlier, digital controllers are good examples of circuits that can be efficiently implemented when modeled as state machines. In the present example, we want to design a TLC with the characteristics summarized in the table of figure 8.10, that is:

· Three modes of operation: Regular, Test, and Standby.

· Regular mode: four states, each with an independent, programmable time, passed to the circuit by means of a CONSTANT.

· Test mode: allows all pre-programmed times to be overwritten (by a manual switch) with a small value, such that the system can be easily tested during maintenance (1 second per state). This value should also be programmable and passed to the circuit using a CONSTANT.

· Standby mode: if set (by a sensor accusing malfunctioning, for example, or a manual switch) the system should activate the yellow lights in both directions and remain so while the standby signal is active.

· Assume that a 60 Hz clock (obtained from the power line itself) is available.

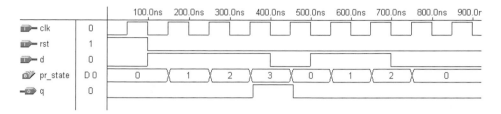

Figure 8.9
Simulation results of example 8.4.

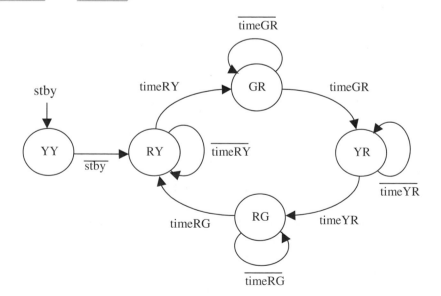

State	Operation Mode		
	REGULAR	TEST	STANDBY
	Time	Time	Time
RG	timeRG (30s)	timeTEST (1s)	---
RY	timeRY (5s)	timeTEST (1s)	---
GR	timeGR (45s)	timeTEST (1s)	---
YR	timeYR (5s)	timeTEST (1s)	---
YY	---	---	Indefinite

Figure 8.10
Specifications and states diagram (regular mode) for example 8.5.

Here, design style #1 can be employed, as shown in the code below.

```
1   -------------------------------------------------
2   LIBRARY ieee;
3   USE ieee.std_logic_1164.all;
4   -------------------------------------------------
5   ENTITY tlc IS
6      PORT ( clk, stby, test: IN STD_LOGIC;
7              r1, r2, y1, y2, g1, g2: OUT STD_LOGIC);
8   END tlc;
9   -------------------------------------------------
10  ARCHITECTURE behavior OF tlc IS
11     CONSTANT timeMAX : INTEGER := 2700;
12     CONSTANT timeRG : INTEGER := 1800;
13     CONSTANT timeRY : INTEGER := 300;
14     CONSTANT timeGR : INTEGER := 2700;
15     CONSTANT timeYR : INTEGER := 300;
16     CONSTANT timeTEST : INTEGER := 60;
17     TYPE state IS (RG, RY, GR, YR, YY);
18     SIGNAL pr_state, nx_state: state;
19     SIGNAL time : INTEGER RANGE 0 TO timeMAX;
20  BEGIN
21     -------- Lower section of state machine: ----
22     PROCESS (clk, stby)
23        VARIABLE count : INTEGER RANGE 0 TO timeMAX;
24     BEGIN
25        IF (stby='1') THEN
26           pr_state <= YY;
27           count := 0;
28        ELSIF (clk'EVENT AND clk='1') THEN
29           count := count + 1;
30           IF (count = time) THEN
31              pr_state <= nx_state;
32              count := 0;
33           END IF;
34        END IF;
35     END PROCESS;
```

```
36      -------- Upper section of state machine: ----
37      PROCESS (pr_state, test)
38      BEGIN
39        CASE pr_state IS
40           WHEN RG =>
41                r1<='1'; r2<='0'; y1<='0'; y2<='0'; g1<='0'; g2<='1';
42                nx_state <= RY;
43                IF (test='0') THEN time <= timeRG;
44                ELSE time <= timeTEST;
45                END IF;
46           WHEN RY =>
47                r1<='1'; r2<='0'; y1<='0'; y2<='1'; g1<='0'; g2<='0';
48                nx_state <= GR;
49                IF (test='0') THEN time <= timeRY;
50                ELSE time <= timeTEST;
51                END IF;
52           WHEN GR =>
53                r1<='0'; r2<='1'; y1<='0'; y2<='0'; g1<='1'; g2<='0';
54                nx_state <= YR;
55                IF (test='0') THEN time <= timeGR;
56                ELSE time <= timeTEST;
57                END IF;
58           WHEN YR =>
59                r1<='0'; r2<='1'; y1<='1'; y2<='0'; g1<='0'; g2<='0';
60                nx_state <= RG;
61                IF (test='0') THEN time <= timeYR;
62                ELSE time <= timeTEST;
63                END IF;
64           WHEN YY =>
65                r1<='0'; r2<='0'; y1<='1'; y2<='1'; g1<='0'; g2<='0';
66                nx_state <= RY;
67        END CASE;
68     END PROCESS;
69 END behavior;
70 -----------------------------------------------------
```

The expected number of flip-flops required to implement this circuit is 15; three
to store pr_state (the machine has five states, so three bits are needed to encode

them), plus twelve for the counter (it is a 12-bit counter, for it must count up to timeMAX = 2700).

Simulation results are shown in figure 8.11. In order for the results to fit properly in the graphs, we adopted small time values, with all CONSTANTS equal to 3 except timeTEST, which was made equal to 1. Therefore, the system is expected to change state every three clock cycles when in Regular operation, or every clock cycle if in Test mode. These two cases can be observed in the first two graphs of figure 8.11, respectively. The third graph shows the Standby mode being activated. As expected, stby is asynchronous and has higher priority than test, causing the system to stay in state YY (state 4) while stby is active. The test signal, on the other hand, is synchronous, but does not need to wait for the current state timing to finish to be activated, as can be observed in the second graph.

Example 8.6:   Signal Generator

We want to design a circuit that, from a clock signal clk, gives origin to the signal outp shown in figure 8.12(a). Notice that the circuit must operate at both edges (rising and falling) of clk.

To circumvent the two-edge aspect (section 6.9), one alternative is to implement two machines, one that operates exclusively at the positive transition of clk and another that operates exclusively at the negative edge, thus generating the intermediate signals out1 and out2 presented in figure 8.12(b). These signals can then be ANDed to give origin to the desired signal outp. Notice that this circuit has no external inputs (except for clk, of course), so the output can only change when clk changes (synchronous output).

```
1   -------------------------------------------
2   ENTITY signal_gen IS
3      PORT ( clk: IN BIT;
4              outp: OUT BIT);
5   END signal_gen;
6   -------------------------------------------
7   ARCHITECTURE fsm OF signal_gen IS
8      TYPE state IS (one, two, three);
9      SIGNAL pr_state1, nx_state1: state;
10     SIGNAL pr_state2, nx_state2: state;
11     SIGNAL out1, out2: BIT;
12 BEGIN
```

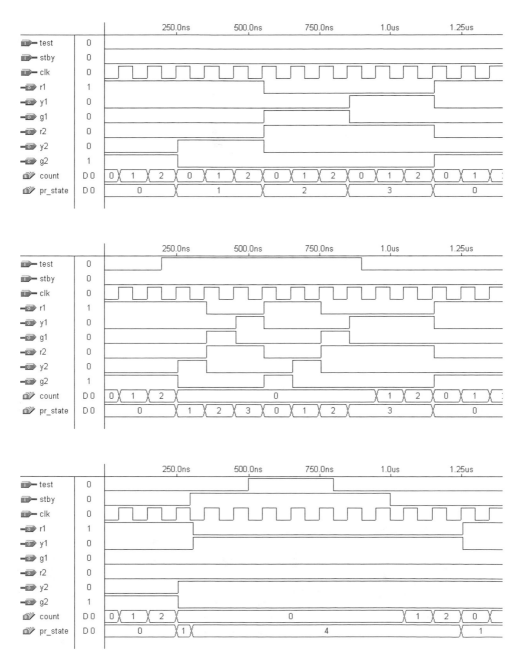

Figure 8.11
Simulation results of example 8.5.

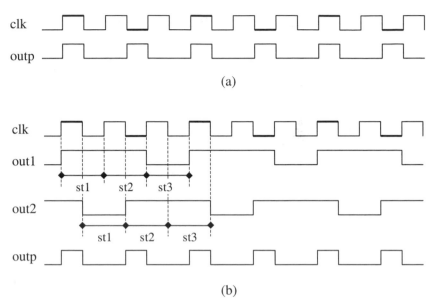

Figure 8.12
Waveforms of example 8.6: (a) signal outp to be generated from clk and (b) intermediate signals out1 and out2 (outp = out1 AND out2).

```
13      ----- Lower section of machine #1: ---
14      PROCESS(clk)
15      BEGIN
16         IF (clk'EVENT AND clk='1') THEN
17             pr_state1 <= nx_state1;
18         END IF;
19      END PROCESS;
20      ----- Lower section of machine #2: ---
21      PROCESS(clk)
22      BEGIN
23         IF (clk'EVENT AND clk='0') THEN
24             pr_state2 <= nx_state2;
25         END IF;
26      END PROCESS;
27      ---- Upper section of machine #1: -----
28      PROCESS (pr_state1)
29      BEGIN
```

```
30          CASE pr_state1 IS
31             WHEN one =>
32                out1 <= '0';
33                nx_state1 <= two;
34             WHEN two =>
35                out1 <= '1';
36                nx_state1 <= three;
37             WHEN three =>
38                out1 <= '1';
39                nx_state1 <= one;
40          END CASE;
41       END PROCESS;
42       ---- Upper section of machine #2: -----
43       PROCESS (pr_state2)
44       BEGIN
45          CASE pr_state2 IS
46             WHEN one =>
47                out2 <= '1';
48                nx_state2 <= two;
49             WHEN two =>
50                out2 <= '0';
51                nx_state2 <= three;
52             WHEN three =>
53                out2 <= '1';
54                nx_state2 <= one;
55          END CASE;
56       END PROCESS;
57       outp <= out1 AND out2;
58 END fsm;
59 -------------------------------------------
```

Simulation results from the circuit synthesized with the code above are shown in figure 8.13.

8.4 Encoding Style: From Binary to OneHot

To encode the states of a state machine, we can select one among several available styles. The default style is *binary*. Its advantage is that it requires the least number of

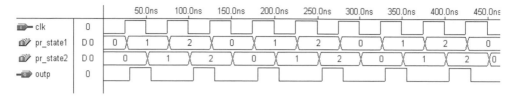

Figure 8.13
Simulation results of example 8.6.

Table 8.1
State encoding of an 8-state FSM.

	Encoding Style		
STATE	BINARY	TWOHOT	ONEHOT
state0	000	00011	00000001
state1	001	00101	00000010
state2	010	01001	00000100
state3	011	10001	00001000
state4	100	00110	00010000
state5	101	01010	00100000
state6	110	10010	01000000
state7	111	01100	10000000

flip-flops. In this case, with n flip-flops (n bits), up to 2^n states can be encoded. The disadvantage of this encoding scheme is that it requires more logic and is slower than the others.

At the other extreme is the *onehot* encoding style, which uses one flip-flop per state. Therefore, it demands the largest number of flip-flops. In this case, with n flip-flops (n bits), only n states can be encoded. On the other hand, this approach requires the least amount of extra logic and is the fastest.

An style that is inbetween the two styles above is the *twohot* encoding scheme, which presents two bits active per state. Therefore, with n flip-flops (n bits), up to $n(n-1)/2$ states can be encoded.

The onehot style is recommended in applications where flip-flops are abundant, like in FPGAs (Field Programmable Gate Arrays). On the other hand, in ASICs (Application Specific Integrated Circuits) the binary style is generally preferred.

As an example, say that our state machine has eight states. Then the encoding would be that shown table 8.1. The number of flip-flops required in each case is three (for binary), five (twohot), or eight (onehot). Other details are also presented in the table.

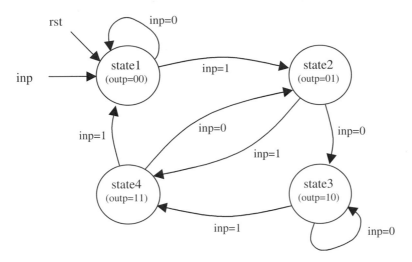

Figure P8.1

8.5 Problems

Each solution to the problems proposed below should be accompanied by synthesis and simulation results. Verify, at least, the following: number of flip-flops inferred and circuit functionality.

Problem 8.1: FSM

Write a VHDL code that implements the FSM described by the states diagram of figure P8.1.

Problem 8.2: Signal Generator #1

Using the FSM approach, design a circuit capable of generating the two signals depicted in figure P8.2 (out1, out2) from a clock signal clk. The signals are periodic and have the same period. However, while one changes only at the rising edge of clk, the other has changes at both edges.

Problem 8.3: Signal Generator #2

Design a finite state machine capable of generating two signals, UP and DOWN, as illustrated in figure P8.3. These signals are controlled by two inputs, GO and STOP. When GO changes from '0' to '1', the output UP must go to '1' too, but $T = 10$ ms later. If GO returns to '0', then UP must return to '0' immediately. However, the output DOWN must now go to '1', again 10 ms later, returning to '0' immediately if

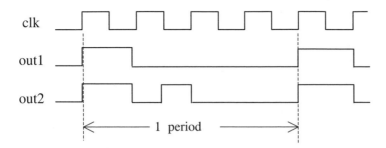

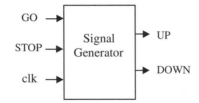

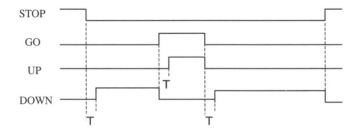

Figure P8.2

Figure P8.3

GO changes to '1'. If the input STOP is asserted, then both outputs must go to '0' immediately and unconditionally. Assume that a 10 kHz clock is available.

Problem 8.4:   Keypad Debouncer and Encoder

Consider the keypad shown in the diagram of figure P8.4. A common way of reading a key press is by means of a technique called *scanning* or *polling*, which reduces the number of wires needed to interconnect the keypad to the main circuit. It consists of sending one column low at a time, while reading each row sequentially. If a key is pressed, then the corresponding row will be low, while the others remain high (due to the pull-up resistors).

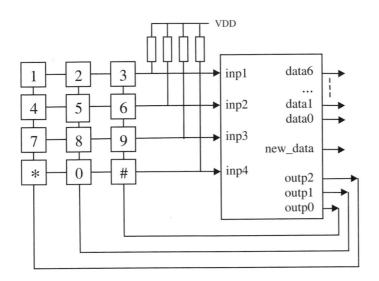

outp	inp	digit	data ASCII
011	0111	1	31h
	1011	4	34h
	1101	7	37h
	1110	*	2Ah
101	0111	2	32h
	1011	5	35h
	1101	8	38h
	1110	0	30h
110	0111	3	33h
	1011	6	36h
	1101	9	39h
	1110	#	23h

Figure P8.4

Encoding: Each digit must be encoded using the ASCII code (7 bits, with the corresponding hexadecimal values listed in the table of figure P8.4). When a new reading is available at the output, the new_data bit should be set to '1'. This will avoid interpreting a key pressed for a long time as a long series of the same character.

Debouncing: A problem inherent to mechanical switches is switch bounces, which occur before a firm contact is finally established. The settling generally takes up to a few milliseconds. Therefore, the choice of the clock frequency is very important. You are asked to choose it such that at least three readings occur in a 5 ms interval. Thus the new_data bit should be turned high only when the same result is obtained in all consecutive readings within a 5 ms interval.

Problem 8.5:   Traffic Light Controller

Using your synthesis tool plus a CPLD/FPGA development kit, implement the TLC of example 8.5. Verify, in the report files generated by your software, which pins of the chip were assigned to the inputs (clk, stby, test) and to the outputs (r1, y1, g1, r2, y2, g2). Then make the following physical connections in your board:

• a 60 Hz square wave signal (from a signal generator), with the appropriate logic levels, to the clk pin.

• a VDD/GND switch to pin stby.

• a VDD/GND switch to pin test.

• an LED (red, if possible), with a 330-1kohm series resistor, to pin r1 (resistor connected between r1 and the anode of the LED, and cathode connected to GND).

• an LED (yellow, if possible) to pin y1, with a series resistor like above.

• an LED (green, if possible) to pin g1, with a series resistor like above.

• finally, install other 3 LEDs, like those above, for r2, y2, and g2.

Now download the program file from your PC to the development kit and verify the operation of the TLC. Play with the switches in order to test all modes of operation. You can also increase the clock frequency to speed up the transition from red to yellow, etc.

Problem 8.6:   Signal Generator #3

Solve problem 8.2 without using the finite state machine approach.

Problem 8.7:   Signal Generator #4

Solve example 8.6 without using the FSM approach.

For further work is this area, see problems 9.3, 9.4, and 9.6 of chapter 9.

9 Additional Circuit Designs

In the preceding chapters, we saw a series of complete design examples utilizing VHDL code. Each design included:

- Top-level diagram of the circuit, with description;
- Review of basic concepts whenever necessary;
- Complete VHDL code;
- Simulation results; and
- Additional comments when needed.

This chapter concludes Part I of the book. In it, a series of additional design examples are presented. These examples, like all the other designs shown so far, are also at the circuit level (that is, self-contained in the main code). In Part II, we will do the same; that is, we will conclude Part II with a chapter containing additional system design examples.

The designs presented in this chapter are the following:

- Barrel shifter (section 9.1)
- Signed and unsigned comparators (section 9.2)
- Carry ripple and carry look ahead adders (section 9.3)
- Fixed-point division (section 9.4)
- Vending machine controller (section 9.5)
- Serial data receiver (section 9.6)
- Parallel-to-serial converter (section 9.7)
- Playing with a SSD (section 9.8)
- Signal generators (section 9.9)
- Memories (section 9.10)
- Finally, a list of problems is also included (section 9.11).

Note: A complete list of all designs presented in the book is shown in section 1.5.

9.1 Barrel Shifter

The diagram of a barrel shifter is shown in figure 9.1. The input is an 8-bit vector. The output is a shifted version of the input, with the amount of shift defined by the "shift" input (from 0 to 7). The circuit consists of three individual barrel shifters, each similar to that seen in example 6.9. Notice that the first barrel has only one '0'

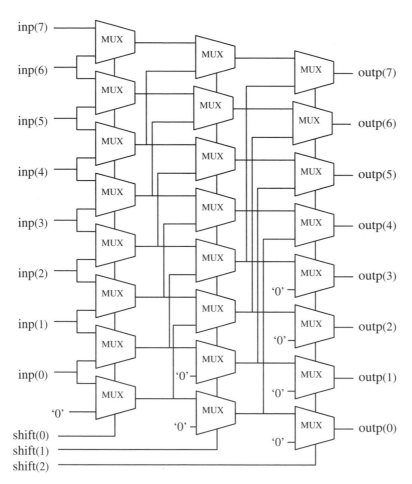

Figure 9.1
Barrel shifter.

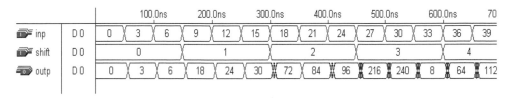

Figure 9.2
Simulation results from barrel shifter of figure 9.1.

connected to one of the multiplexers (bottom left corner), while the second has two, and the third has four. For larger vectors, we would just keep doubling the number of '0' inputs. If shift = "001", for example, then only the first barrel should cause a shift; on the other hand, if shift = "111", then all barrels should cause a shift.

A VHDL code for the circuit of figure 9.1 is presented below. Simulation results, verifying the functionality of the circuit, are shown in figure 9.2. As can be seen in the latter, the output is equal to the input when shift = 0 (that is, shift = "000"). It can also be seen that, as long as no bit of value '1' is shifted out of the barrel, the output is equal to the input multiplied by 2 (1 shift) when shift = 1 ("001"), multiplied by 4 (2 shifts) when shift = 2 ("010"), multiplied by 8 (3 shifts) when shift = 3 ("011"), and so on.

```
1   ------------------------------------------
2   LIBRARY ieee;
3   USE ieee.std_logic_1164.all;
4   ------------------------------------------
5   ENTITY barrel IS
6      PORT ( inp: IN STD_LOGIC_VECTOR (7 DOWNTO 0);
7             shift: IN STD_LOGIC_VECTOR (2 DOWNTO 0);
8             outp: OUT STD_LOGIC_VECTOR (7 DOWNTO 0));
9   END barrel;
10  ------------------------------------------
11  ARCHITECTURE behavior OF barrel IS
12  BEGIN
13     PROCESS (inp, shift)
14        VARIABLE temp1: STD_LOGIC_VECTOR (7 DOWNTO 0);
15        VARIABLE temp2: STD_LOGIC_VECTOR (7 DOWNTO 0);
16     BEGIN
```

```
17          ---- 1st shifter -----
18          IF (shift(0)='0') THEN
19              temp1 := inp;
20          ELSE
21              temp1(0) := '0';
22              FOR i IN 1 TO inp'HIGH LOOP
23                  temp1(i) := inp(i-1);
24              END LOOP;
25          END IF;
26          ---- 2nd shifter -----
27          IF (shift(1)='0') THEN
28              temp2 := temp1;
29          ELSE
30              FOR i IN 0 TO 1 LOOP
31                  temp2(i) := '0';
32              END LOOP;
33              FOR i IN 2 TO inp'HIGH LOOP
34                  temp2(i) := temp1(i-2);
35              END LOOP;
36          END IF;
37          ---- 3rd shifter -----
38          IF (shift(2)='0') THEN
39              outp <= temp2;
40          ELSE
41              FOR i IN 0 TO 3 LOOP
42                  outp(i) <= '0';
43              END LOOP;
44              FOR i IN 4 TO inp'HIGH LOOP
45                  outp(i) <= temp2(i-4);
46              END LOOP;
47          END IF;
48      END PROCESS;
49 END behavior;
50 ---------------------------------------------
```

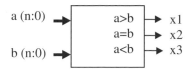

Figure 9.3
Comparator.

9.2 Signed and Unsigned Comparators

Figure 9.3 shows the top-level diagram of a comparator. The size of the vectors to be compared is generic (n + 1). Three outputs must be provided: one corresponding to a > b, another to a = b, and finally one relative to a < b. Three solutions are presented: the first considers a and b as signed numbers, while the other two consider them as unsigned values. Simulation results are also included.

Signed Comparator

Notice the presence of the std_logic_arith package in the code below (line 4), which is necessary to operate with SIGNED (or UNSIGNED) data types (a and b were declared as SIGNED numbers in line 8).

```
1    ---- Signed Comparator: ----------------
2    LIBRARY ieee;
3    USE ieee.std_logic_1164.all;
4    USE ieee.std_logic_arith.all;   -- necessary!
5    ----------------------------------------
6    ENTITY comparator IS
7       GENERIC (n: INTEGER := 7);
8       PORT (a, b: IN SIGNED (n DOWNTO 0);
9             x1, x2, x3: OUT STD_LOGIC);
10   END comparator;
11   ----------------------------------------
12   ARCHITECTURE signed OF comparator IS
13   BEGIN
14      x1 <= '1' WHEN a > b ELSE '0';
15      x2 <= '1' WHEN a = b ELSE '0';
16      x3 <= '1' WHEN a < b ELSE '0';
17   END signed;
18   ----------------------------------------
```

Figure 9.4
Simulation result of *signed* comparator of figure 9.3.

Simulation results are shown in figure 9.4. As can be seen, $127 > 0$, but $128 < 0$ and also $255 < 0$ (because in 2's complement notation 127 is the decimal 127 itself, but 128 is the decimal -128, and 255 is indeed -1).

Unsigned Comparator #1

The VHDL code below is the counterpart of the code just presented (signed comparator). Notice again the presence of the std_logic_arith package (line 4), which is necessary to operate with UNSIGNED (or SIGNED) data types (a and b were declared as UNSIGNED numbers in line 8).

```
1  ---- Unsigned Comparator #1: -----------
2  LIBRARY ieee;
3  USE ieee.std_logic_1164.all;
4  USE ieee.std_logic_arith.all;  -- necessary!
5  -----------------------------------------
6  ENTITY comparator IS
7     GENERIC (n: INTEGER := 7);
8     PORT (a, b: IN UNSIGNED (n DOWNTO 0);
9           x1, x2, x3: OUT STD_LOGIC);
10 END comparator;
11 -----------------------------------------
12 ARCHITECTURE unsigned OF comparator IS
13 BEGIN
14    x1 <= '1' WHEN a > b ELSE '0';
15    x2 <= '1' WHEN a = b ELSE '0';
16    x3 <= '1' WHEN a < b ELSE '0';
17 END unsigned;
18 -----------------------------------------
```

Figure 9.5
Simulation result of *unsigned* comparator of figure 9.3.

Unsigned Comparator #2

Unsigned comparators can also be implemented with STD_LOGIC_VECTORS, in which case there is no need to declare the std_logic_arith package. A solution of this kind is presented below.

```
1   ---- Unsigned Comparator #2: -----------
2   LIBRARY ieee;
3   USE ieee.std_logic_1164.all;
4   ----------------------------------------
5   ENTITY comparator IS
6      GENERIC (n: INTEGER := 7);
7      PORT (a, b: IN STD_LOGIC_VECTOR (n DOWNTO 0);
8            x1, x2, x3: OUT STD_LOGIC);
9   END comparator;
10  ----------------------------------------
11  ARCHITECTURE unsigned OF comparator IS
12  BEGIN
13     x1 <= '1' WHEN a > b ELSE '0';
14     x2 <= '1' WHEN a = b ELSE '0';
15     x3 <= '1' WHEN a < b ELSE '0';
16  END unsigned;
17  ----------------------------------------
```

Simulation results (from either unsigned comparator) are shown in figure 9.5. Contrary to figure 9.4, now 128 and 255 are indeed bigger than zero.

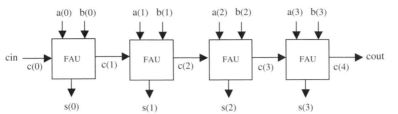

a b	cin	s	cout
0 0	0	0	0
0 1	0	1	0
1 0	0	1	0
1 1	0	0	1
0 0	1	1	0
0 1	1	0	1
1 0	1	0	1
1 1	1	1	1

Figure 9.6
4-bit carry-ripple adder and truth table of Full-Adder Unit (FAU).

9.3 Carry-Ripple and Carry-Lookahead Adders

Carry-ripple and carry-lookahead are two classical approaches to the design of adders. The former has the advantage of requiring less hardware (with restrictions–see comments in example 6.8), while the latter is faster. Both approaches are discussed below.

Carry-Ripple Adder

Figure 9.6 shows a 4-bit unsigned carry-ripple adder. For each bit, a Full-Adder Unit (FAU, section 1.4) is employed. The truth table of the FAU is also shown. In it, a and b represent the input bits, cin is the carry-in bit, s is the sum bit, and cout is the carry-out bit. s must be high whenever the number of inputs that are high is odd (parity function), while cout must be high when two or more inputs are high (majority function). Notice in figure 9.6 that each FAU relies on the carry bit produced by the previous stage. This approach minimizes the size of the circuitry, at the expense of increased propagation delay.

Based on the truth table of figure 9.6, a very simple way of computing s and cout is the following:

s = a XOR b XOR cin

cout = (a AND b) OR (a AND cin) OR (b AND cin)

Therefore, a VHDL implementation of the carry ripple adder is straightforward. The solution shown below works for any number (n) of input bits, defined by means of a GENERIC statement in line 5. Simulation results from the circuit synthesized with the code below are shown in figure 9.7.

```
1   LIBRARY ieee;
2   USE ieee.std_logic_1164.all;
```

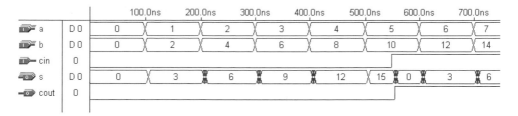

Figure 9.7
Simulation results from the carry-ripple adder of figure 9.6.

```
3   ----------------------------------------------
4   ENTITY adder_cripple IS
5      GENERIC (n: INTEGER := 4);
6      PORT ( a, b: IN STD_LOGIC_VECTOR (n-1 DOWNTO 0);
7             cin: IN STD_LOGIC;
8             s: OUT STD_LOGIC_VECTOR (n-1 DOWNTO 0);
9             cout: OUT STD_LOGIC);
10  END adder_cripple;
11  ----------------------------------------------
12  ARCHITECTURE adder OF adder_cripple IS
13     SIGNAL c: STD_LOGIC_VECTOR (n DOWNTO 0);
14  BEGIN
15     c(0) <= cin;
16     G1: FOR i IN 0 TO n-1 GENERATE
17        s(i) <= a(i) XOR b(i) XOR c(i);
18        c(i+1) <= (a(i) AND b(i)) OR
19                  (a(i) AND c(i)) OR
20                  (b(i) AND c(i));
21     END GENERATE;
22     cout <= c(n);
23  END adder;
24  ----------------------------------------------
```

Pre-defined "+" Operator

We have already seen in example 6.8 that an adder can be implemented directly with
the "+" operator (section 4.1). In this case, a circuit optimized for the target technology will normally be implemented by the compiler. If, nevertheless, one wants the
solution to be of a specific type, then an explicit code must be written.

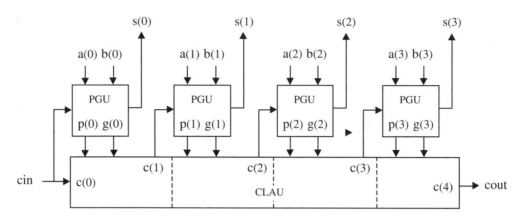

Figure 9.8
4-bit carry look ahead adder.

Carry-Lookahead Adder

A diagram of a 4-bit carry-lookahead adder is shown in figure 9.8. Its implementation is based on the *generate* and *propagate* concept, which gives the circuit higher speed than its carry-ripple adder counterpart (at the expense of more silicon area).

Consider two input bits, a and b. The *generate* (g) and *propagate* (p) signals are defined as:

$g = a$ AND b

$p = a$ XOR b

Notice that such signals can be computed in advance, because neither depends on the carry bit.

If we consider now two input vectors, $a = a(n-1) \ldots a(1)a(0)$ and $b = b(n-1) \ldots b(1)b(0)$, then the corresponding generate and propagate vectors are $g = g(n-1) \ldots g(1)g(0)$ and $p = p(n-1) \ldots p(1)p(0)$, where

$g(j) = a(j)$ AND $b(j)$

$p(j) = a(j)$ XOR $b(j)$

Let us consider now the carry vector, $c = c(n-1) \ldots c(1)c(0)$. The carry bits can be computed from g and p:

$c(0) \equiv cin$

$c(1) = c(0)p(0) + g(0)$

$$c(2) = c(0)p(0)p(1) + g(0)p(1) + g(1)$$

$$c(3) = c(0)p(0)p(1)p(2) + g(0)p(1)p(2) + g(1)p(2) + g(2), \text{ etc.}$$

Hence, contrary to the carry-ripple adder, each carry bit above is computed independently; that is, none of the expressions above depends on preceding carry computations, and that is the reason why this circuit is faster. On the other hand, the hardware complexity grows very fast, limiting this approach to just a few bits (typically four). Larger adders can be implemented by associating such 4-bit-or-so units.

The implementation of the adder of figure 9.8 is now straightforward. The PGU (Propagate-Generate Unit) computes p and g (four units are required), plus the actual sum (s), while the CLAU (Carry-Lookahead Unit) computes the carry bits.

Note: In order to construct bigger carry-lookahead adders, the CLAU block of figure 9.8 must posses Group Propagate (GP) and Group Generate (GG) outputs, which were omitted in the figure because this implementation is intended for four bits only.

```
1  ------------------------------------------------
2  LIBRARY ieee;
3  USE ieee.std_logic_1164.all;
4  ------------------------------------------------
5  ENTITY CLA_Adder IS
6     PORT ( a, b: IN STD_LOGIC_VECTOR (3 DOWNTO 0);
7            cin: IN STD_LOGIC;
8            s: OUT STD_LOGIC_VECTOR (3 DOWNTO 0);
9            cout: OUT STD_LOGIC);
10 END CLA_Adder;
11 ------------------------------------------------
12 ARCHITECTURE CLA_Adder OF CLA_Adder IS
13    SIGNAL c: STD_LOGIC_VECTOR (4 DOWNTO 0);
14    SIGNAL p: STD_LOGIC_VECTOR (3 DOWNTO 0);
15    SIGNAL g: STD_LOGIC_VECTOR (3 DOWNTO 0);
16 BEGIN
17    ---- PGU: -------------------------------
18    G1: FOR i IN 0 TO 3 GENERATE
19       p(i) <= a(i) XOR b(i);
20       g(i) <= a(i) AND b(i);
21       s(i) <= p(i) XOR c(i);
```

```
22    END GENERATE;
23    ---- CLAU: ------------------------------
24    c(0) <= cin;
25    c(1) <= (cin AND p(0)) OR
26             g(0);
27    c(2) <= (cin AND p(0) AND p(1)) OR
28             (g(0) AND p(1)) OR
29             g(1);
30    c(3) <= (cin AND p(0) AND p(1) AND p(2)) OR
31             (g(0) AND p(1) AND p(2)) OR
32             (g(1) AND p(2)) OR
33             g(2);
34    c(4) <= (cin AND p(0) AND p(1) AND p(2) AND p(3)) OR
35             (g(0) AND p(1) AND p(2) AND p(3)) OR
36             (g(1) AND p(2) AND p(3)) OR
37             (g(2) AND p(3)) OR
38             g(3);
39    cout <= c(4);
40 END CLA_Adder;
41 ------------------------------------------
```

Qualitatively, the simulation results obtained from the circuit synthesized with the code above are similar to those from the carry-ripple adder presented in figure 9.7.

9.4 Fixed-Point Division

We saw in chapter 4 that the pre-defined "/" (division) operator accepts only power of two divisors, that is, it is indeed a "shift" operator. In this section, we will discuss the implementation of *generic* division, in which the dividend and divisor can be any integer. We start by describing the division algorithm, then we present two VHDL solutions followed by simulation results.

Division Algorithm

Say that we want to calculate $y = a/b$, where a, b, and y have the same number $(n + 1)$ of bits. The algorithm is illustrated in figure 9.9, for a = "1011" (decimal 11) and b = "0011" (decimal 3), from which we expect y = "0011" (decimal 3) and remainder "0010" (decimal 2). We first create a shifted version of b, whose length is $2n + 1$ bits (shown in the b-related column in figure 9.9). b_inp(i) is simply b shifted to the left by i positions (notice the underscored characters in the b-related column).

Index (i)	a-related input (a_inp)	Comparison	b-related input (b_inp)	y (quotient)	Operation on 1st column
3	1011	<	0011000	0	none
2	1011	<	0001100	0	none
1	1011	>	0000110	1	a_inp(i)-b_inp(i)
0	0101	>	0000011	1	a_inp(i)-b_inp(i)
	0010 (rem)				

Figure 9.9
Division algorithm.

The computation of the quotient is performed as follows. Starting from the top of the table, we compare a_inp(i) with b_inp(i). If the former is bigger than or equal to the latter, than y(i) = '1' and b_inp(i) is subtracted from a_inp(i); otherwise, y(i) = '0' and we simply proceed to the next line. After $n + 1$ iterations, the computation is completed and the value left in a_inp is the remainder.

Note: It is obvious that, to subtract b_inp from a_inp, the number of bits of a_inp cannot be less than that of b_inp, so the actual length of a_inp must be increased, which is attained by simply filling a_inp with n '0's on its left-hand side ('0's not shown in figure 9.9).

Another way of presenting the division algorithm is the following. We multiply b by $2^{**}n$, where $n + 1$ is the number of bits. This, of course, corresponds to shifting b n positions to the left, but without throwing out any of its bits (so the new b-vector must be n bits longer than the original vector). If a is bigger than the new b, then y(n) = '1', and b (the new value) must be subtracted from a; otherwise, y(n) = '0'. Now we move to the next iteration. We multiply b (the original value) by $2^{**}(n - 1)$, which is equivalent to shifting the original vector $n - 1$ positions to the left, or shifting the value of b just used in the previous computation back one position to the right. Then we compare it to a, as we did before, to decide whether $y(n - 1)$ should be '1' or '0', and so on.

VHDL Dividers

Below are two solutions for the division problem. Both use sequential code: IF is used in the first, while LOOP plus IF are employed in the second. The first solution is a step-by-step code, so the division algorithm described above can be clearly observed. The second is more compact and is also generic (notice that n was defined

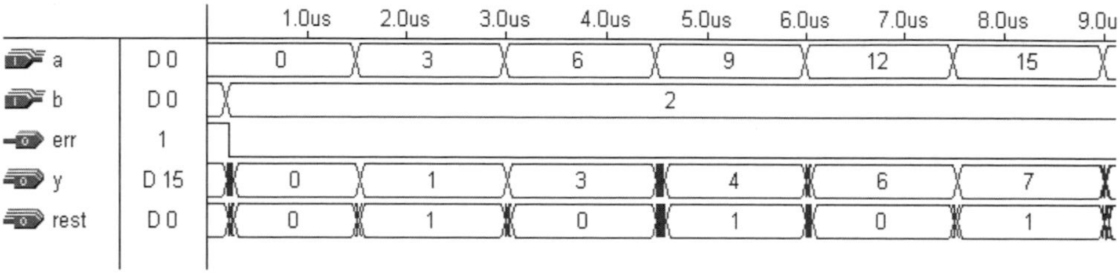

Figure 9.10
Simulation results of divider (for 4-bit operands).

by means of a GENERIC statement in line 6). The solutions include also a b = 0 check routine.

Simulation results are shown in figure 9.10.

```
1   ----- Solution 1: step-by-step --------------------
2   LIBRARY ieee;
3   USE ieee.std_logic_1164.all;
4   ----------------------------------------------------
5   ENTITY divider IS
6      PORT ( a, b: IN INTEGER RANGE 0 TO 15;
7              y: OUT STD_LOGIC_VECTOR (3 DOWNTO 0);
8              rest: OUT INTEGER RANGE 0 TO 15;
9              err : OUT STD_LOGIC);
10  END divider;
11  ----------------------------------------------------
12  ARCHITECTURE rtl OF divider IS
13  BEGIN
14     PROCESS (a, b)
15        VARIABLE temp1: INTEGER RANGE 0 TO 15;
16        VARIABLE temp2: INTEGER RANGE 0 TO 15;
17     BEGIN
18        ----- Error and initialization: -------
19        temp1 := a;
20        temp2 := b;
21        IF (b=0) THEN err <= '1';
22        ELSE err <= '0';
23        END IF;
24        ----- y(3): --------------------------
```

```
25        IF (temp1 >= temp2 * 8) THEN
26            y(3) <= '1';
27            temp1 := temp1 - temp2*8;
28        ELSE y(3) <= '0';
29        END IF;
30        ----- y(2): --------------------------
31        IF (temp1 >= temp2 * 4) THEN
32            y(2) <= '1';
33            temp1 := temp1 - temp2 * 4;
34        ELSE y(2) <= '0';
35        END IF;
36        ----- y(1): --------------------------
37        IF (temp1 >= temp2 * 2) THEN
38            y(1) <= '1';
39            temp1 := temp1 - temp2 * 2;
40        ELSE y(1) <= '0';
41        END IF;
42        ----- y(0): --------------------------
43        IF (temp1 >= temp2) THEN
44            y(0) <= '1';
45            temp1 := temp1 - temp2;
46        ELSE y(0) <= '0';
47        END IF;
48        ----- Remainder: ---------------------
49        rest <= temp1;
50     END PROCESS;
51 END rtl;
52 ----------------------------------------------------

1  ------ Solution 2: compact and generic -----------
2  LIBRARY ieee;
3  USE ieee.std_logic_1164.all;
4  ----------------------------------------------------
5  ENTITY divider IS
6     GENERIC(n: INTEGER := 3);
7     PORT ( a, b: IN INTEGER RANGE 0 TO 15;
8             y: OUT STD_LOGIC_VECTOR (3 DOWNTO 0);
9             rest: OUT INTEGER RANGE 0 TO 15;
```

```
10              err : OUT STD_LOGIC);
11 END divider;
12 ---------------------------------------------------
13 ARCHITECTURE rtl OF divider IS
14 BEGIN
15    PROCESS (a, b)
16       VARIABLE temp1: INTEGER RANGE 0 TO 15;
17       VARIABLE temp2: INTEGER RANGE 0 TO 15;
18    BEGIN
19       ----- Error and initialization: -------
20       temp1 := a;
21       temp2 := b;
22       IF (b=0) THEN err <= '1';
23       ELSE err <= '0';
24       END IF;
25       ----- y: ---------------------------
26       FOR i IN n DOWNTO 0 LOOP
27          IF(temp1 >= temp2 * 2**i) THEN
28             y(i) <= '1';
29             temp1 := temp1 - temp2 * 2**I;
30          ELSE y(i) <= '0';
31          END IF;
32       END LOOP;
33       ----- Remainder: ---------------------
34       rest <= temp1;
35    END PROCESS;
36 END rtl;
37 ---------------------------------------------------
```

9.5   Vending-Machine Controller

In this example, we will design a controller for a vending machine, which sells candy bars for twenty-five cents. As seen in chapter 8, this is the type of design where the FSM (finite state machine) model is helpful.

The inputs and outputs of the controller are shown in figure 9.11. The input signals nickel_in, dime_in, and quarter_in indicate that a corresponding coin has been deposited. Two additional inputs, clk (clock) and rst (reset), are also necessary. The

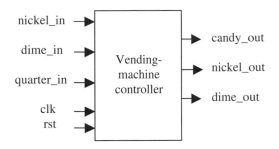

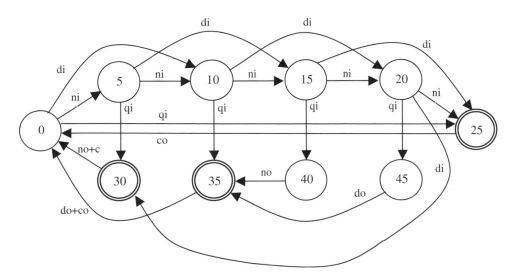

Figure 9.11
Vending-machine controller (top-level and states diagrams). The signals are. ni = nickel_in, di = dime_in, qi = quarter_in, no = nickel_out, do = dime_out, and co = candy_out.

controller responds with three outputs: candy_out, to dispense a candy bar, plus nickel_out and dime_out, asserted when change is due.

Figure 9.11 also shows the states of the corresponding FSM. The numbers inside the circles represent the total amount deposited by the customer (only nickels, dimes, and quarters are accepted). State 0 is the idle state. From it, if a nickel is deposited, the machine moves to state 5; if a dime, to state 10; or if a quarter, to state 25. Similar situations are repeated for all states, up to state 20. If state 25 is reached, then a candy bar is dispensed, with no change. However, if state 40 is reached, for example, then a nickel is delivered, passing therefore the system to state 35, from which a dime is delivered and a candy bar dispensed. The three states marked with double circles are those from which a candy bar is delivered and the machine returns to state 0.

This problem will be divided into two parts: in the first, the *fundamental* aspects related to the design of the vending machine controller (figure 9.11) are treated; in the second, additional (and *indispensable*) features are added. The first part is studied in this section, while the second is proposed as a problem (problem 9.3). The introduction of such additional features is necessary for safety reasons; since we are dealing with money, we must assure that none of the parts (machine or customer) will be hurt in the transaction.

A VHDL code, treating only the basic features of the problem depicted in figure 9.11, is presented below. We have assumed that the additional features proposed in problem 9.3 will indeed be implemented, in which case glitches are acceptable in the first part of the solution. Therefore, design style #1 (section 8.2) can be employed.

The enumerated type *state* (line 12) contains a list of all states shown in the FSM diagram of figure 9.11. There are ten states, so four bits are necessary to encode them (so four flip-flops will be inferred). Recall that the compiler encodes such states in the order that they are listed, so st0 = "0000" (decimal 0), st5 = "0001" (decimal 1), ..., st45 = "1001" (decimal 9). Therefore, in the simulations, such numbers are shown instead of the state names.

```
1     ------------------------------------------------------
2     LIBRARY ieee;
3     USE ieee.std_logic_1164.all;
4     ------------------------------------------------------
5     ENTITY vending_machine IS
6        PORT ( clk, rst: IN STD_LOGIC;
7                 nickel_in, dime_in, quarter_in: IN BOOLEAN;
8                 candy_out, nickel_out, dime_out: OUT STD_LOGIC);
9     END vending_machine;
```

```
10   --------------------------------------------------------
11   ARCHITECTURE fsm OF vending_machine IS
12      TYPE state IS (st0, st5, st10, st15, st20, st25,
13         st30, st35, st40, st45);
14      SIGNAL present_state, next_state: STATE;
15   BEGIN
16      ---- Lower section of the FSM (Sec. 8.2): ---------
17      PROCESS (rst, clk)
18      BEGIN
19         IF (rst='1') THEN
20            present_state <= st0;
21         ELSIF (clk'EVENT AND clk='1') THEN
22            present_state <= next_state;
23         END IF;
24      END PROCESS;
25      ---- Upper section of the FSM (Sec. 8.2): ---------
26      PROCESS (present_state, nickel_in, dime_in, quarter_in)
27      BEGIN
28         CASE present_state IS
29            WHEN st0 =>
30               candy_out <= '0';
31               nickel_out <= '0';
32               dime_out <= '0';
33               IF (nickel_in) THEN next_state <= st5;
34               ELSIF (dime_in) THEN next_state <= st10;
35               ELSIF (quarter_in) THEN next_state <= st25;
36               ELSE next_state <= st0;
37               END IF;
38            WHEN st5 =>
39               candy_out <= '0';
40               nickel_out <= '0';
41               dime_out <= '0';
42               IF (nickel_in) THEN next_state <= st10;
43               ELSIF (dime_in) THEN next_state <= st15;
44               ELSIF (quarter_in) THEN next_state <= st30;
45               ELSE next_state <= st5;
46               END IF;
```

```
47        WHEN st10 =>
48            candy_out <= '0';
49            nickel_out <= '0';
50            dime_out <= '0';
51            IF (nickel_in) THEN next_state <= st15;
52            ELSIF (dime_in) THEN next_state <= st20;
53            ELSIF (quarter_in) THEN next_state <= st35;
54            ELSE next_state <= st10;
55            END IF;
56        WHEN st15 =>
57            candy_out <= '0';
58            nickel_out <= '0';
59            dime_out <= '0';
60            IF (nickel_in) THEN next_state <= st20;
61            ELSIF (dime_in) THEN next_state <= st25;
62            ELSIF (quarter_in) THEN next_state <= st40;
63            ELSE next_state <= st15;
64            END IF;
65        WHEN st20 =>
66            candy_out <= '0';
67            nickel_out <= '0';
68            dime_out <= '0';
69            IF (nickel_in) THEN next_state <= st25;
70            ELSIF (dime_in) THEN next_state <= st30;
71            ELSIF (quarter_in) THEN next_state <= st45;
72            ELSE next_state <= st20;
73            END IF;
74        WHEN st25 =>
75            candy_out <= '1';
76            nickel_out <= '0';
77            dime_out <= '0';
78            next_state <= st0;
79        WHEN st30 =>
80            candy_out <= '1';
81            nickel_out <= '1';
82            dime_out <= '0';
83            next_state <= st0;
```

```
84                WHEN st35 =>
85                    candy_out <= '1';
86                    nickel_out <= '0';
87                    dime_out <= '1';
88                    next_state <= st0;
89                WHEN st40 =>
90                    candy_out <= '0';
91                    nickel_out <= '1';
92                    dime_out <= '0';
93                    next_state <= st35;
94                WHEN st45 =>
95                    candy_out <= '0';
96                    nickel_out <= '0';
97                    dime_out <= '1';
98                    next_state <= st35;
99        END CASE;
100       END PROCESS;
101
102       END fsm;
103       -----------------------------------------------------
```

Simulation results are presented in figure 9.12. As can be seen, three nickels and one quarter were deposited. Notice that, at the first positive clock edge after the first nickel was deposited, the FSM moves from state st0 (decimal 0) to st5 (decimal 1);

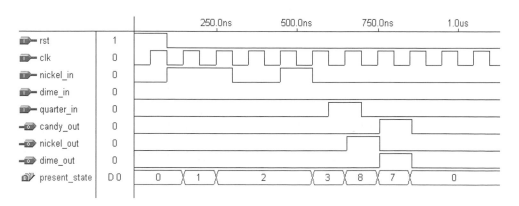

Figure 9.12
Simulation results from the vending-machine controller.

after de second nickel, to state st10 (decimal 2); after de third, to state st15 (decimal 3); and, after de quarter has been deposited, to state st40 (decimal 8). After that, a nickel is returned to the customer (nickel_out = '1'), causing the FSM to move to state st35 (decimal 7), at which a dime is delivered (dime_out = '1') and a candy bar is dispensed (candy_out = '1'). The system returns then to its idle state (st0).

As mentioned above, additional features (like handshake) are necessary to increase the security of the transactions. Please refer to problem 9.3 for a continuation of this design.

9.6 Serial Data Receiver

The diagram of a serial data receiver is shown in figure 9.13. It contains a serial data input, din, and a parallel data output, data(6:0). A clock signal is also needed at the input. Two supervision signals are generated by the circuit: err (error) and data_valid.

The input train consists of ten bits. The first bit is a start bit, which, when high, must cause the circuit to start receiving data. The next seven are the actual data bits. The ninth bit is a parity bit, whose status must be '0' if the number of ones in data is even, or '1' otherwise. Finally, the tenth is a stop bit, which must be high if the transmission is correct. An error is detected when either the parity does not check or the stop bit is not a '1'. When reception is concluded and if no error has been detected, then the data stored in the internal registers (reg) is transferred to data(6:0) and the data_valid output is asserted.

A VHDL code for this circuit is presented below. A few variables were used: count, to determine the number of bits received; reg, which stores the data; and temp, to compute the error. Notice in line 37 that reg(0) = din was used instead of reg(0) = '0', because we want the time slot immediately after the stop bit to be considered as possibly containing a start bit for the next input train.

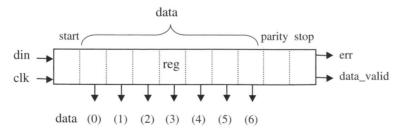

Figure 9.13
Serial data receiver.

```
1       ------------------------------------------------
2       LIBRARY ieee;
3       USE ieee.std_logic_1164.all;
4       ------------------------------------------------
5       ENTITY receiver IS
6          PORT ( din, clk, rst: IN BIT;
7                    data: OUT BIT_VECTOR (6 DOWNTO 0);
8                    err, data_valid: OUT BIT);
9       END receiver;
10 ------------------------------------------------
11 ARCHITECTURE rtl OF receiver IS
12 BEGIN
13    PROCESS (rst, clk)
14       VARIABLE count: INTEGER RANGE 0 TO 10;
15       VARIABLE reg: BIT_VECTOR (10 DOWNTO 0);
16       VARIABLE temp : BIT;
17    BEGIN
18       IF (rst='1') THEN
19          count:=0;
20          reg := (reg'RANGE => '0');
21          temp := '0';
22          err <= '0';
23          data_valid <= '0';
24       ELSIF (clk'EVENT AND clk='1') THEN
25          IF (reg(0)='0' AND din='1') THEN
26             reg(0) := '1';
27          ELSIF (reg(0)='1') THEN
28             count := count + 1;
29             IF (count < 10) THEN
30                reg(count) := din;
31             ELSIF (count = 10) THEN
32                temp := (reg(1) XOR reg(2) XOR reg(3) XOR
33                         reg(4) XOR reg(5) XOR reg(6) XOR
34                         reg(7) XOR reg(8)) OR NOT reg(9);
35                err <= temp;
36                count := 0;
37                reg(0) := din;
38                IF (temp = '0') THEN
```

```
39                    data_valid <= '1';
40                    data <= reg(7 DOWNTO 1);
41                END IF;
42              END IF;
43            END IF;
44        END IF;
45    END PROCESS;
46 END rtl;
47 -------------------------------------------------
```

Simulation results are presented in figure 9.14. The input sequence is din = {start = 1, din = 0111001, parity = 0, stop = 1}. As can be seen in the upper graph, no error was detected in this case, because the parity and stop bits are correct. Hence, after count reaches 9, the data is made available, that is, data = 0111001, from data(0) to data(6), which corresponds to the decimal 78, and the data_valid bit is

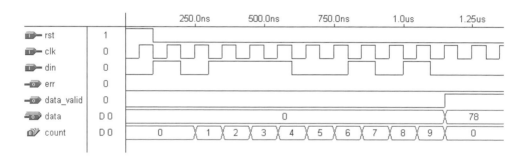

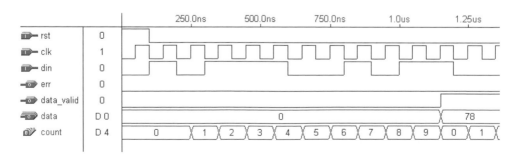

Figure 9.14
Simulation results of serial data receivers.

asserted. Notice that the output remains so indefinitely, unless a new input train is received.

The only difference in the lower graph is that a start bit appears immediately after the stop bit. As can be seen, the count variable starts then to count and the whole process is repeated.

9.7   Parallel-to-Serial Converter

A parallel-to-serial converter is a typical application of shift registers. It consists of sending out a block of data serially. The need for such converters arises, for example, in ASIC chips when there are not enough pins available to output all data bits simultaneously.

A diagram of a parallel-to-serial converter is presented in figure 9.15. d(7:0) is the data vector to be sent out, while dout is the actual output. There are also two other inputs: clk and load. When load is asserted, d is synchronously stored in the shift register reg. While load stays high, the MSB, d(7), remains available at the output. Once load is returned to '0', the subsequent bits are presented at the output at each positive edge of clk. After all eight bits have been sent out, the output remains low until the next transmission.

```
1        --------------------------------------------------
2        LIBRARY ieee;
3        USE ieee.std_logic_1164.all;
4        --------------------------------------------------
5        ENTITY serial_converter IS
6           PORT ( d: IN STD_LOGIC_VECTOR (7 DOWNTO 0);
7                    clk, load: IN STD_LOGIC;
8                    dout: OUT STD_LOGIC);
9        END serial_converter;
10 --------------------------------------------------
```

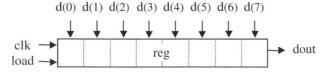

Figure 9.15
Parallel-to-serial converter.

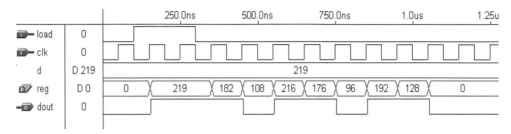

Figure 9.16
Simulation results of parallel-to-serial converter.

```
11 ARCHITECTURE serial_converter OF serial_converter IS
12    SIGNAL reg: STD_LOGIC_VECTOR (7 DOWNTO 0);
13 BEGIN
14    PROCESS (clk)
15    BEGIN
16       IF (clk'EVENT AND clk='1') THEN
17          IF (load='1') THEN  reg <= d;
18          ELSE reg <= reg(6 DOWNTO 0) & '0';
19          END IF;
20       END IF;
21    END PROCESS;
22    dout <= reg(7);
23 END serial_converter;
24 -------------------------------------------------
```

Simulation results from the circuit synthesized with the code above are shown in figure 9.16. d = "11011011" (decimal 219) was chosen. As can be seen, d(7) = '1' is presented at the output at the first rising edge of clk after load has been asserted, staying there while load remains high (to illustrate this fact, load was kept high during two clock cycles). The other bits follow as soon as load returns to '0'. Notice that after all bits have been transmitted, the output stays low.

9.8   Playing with a Seven-Segment Display

We want to design a little game with an SSD (seven-segment display). The top-level diagram of the circuit is shown in figure 9.17. It contains two inputs, clk and stop, and one output, dout(6:0), which feeds the SSD. Assume that $f_{clk} = 1$ kHz.

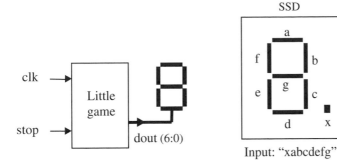

Figure 9.17
Playing with an SSD.

Our circuit should cause a continuous clockwise movement of the SSD segments. Also, in order to make the circulatory movement more realistic, we want to momentarily overlap neighboring segments. Consequently, the sequence should be a → ab → b → bc → c → cd → d → de → e → ef → f → fa → a, with the combined states (ab, bc, etc.) lasting only a few milliseconds. If stop is asserted, then the circuit should return to state a and remain so until stop is turned low again.

From chapter 8, it is clear that this is a circuit for which the FSM approach is appropriate. The states diagram is presented in figure 9.18. We want the system to remain in states a, b, c, etc. for time1 = 80 ms, and in the combined states, ab, bc, etc., for time2 = 30 ms. Therefore, a counter counting up to 80 (the clock period is 1 ms) or up to 30 can be employed to determine when to move to the next state.

A VHDL solution is shown below. Notice that it is a straight implementation of the FSM template seen in section 8.2. In lines 11–12, time1 and time2 were declared as two constants. Small values (4 and 2, respectively) were here used in order for the simulation results to fit well in one plot, but 80 and 30, respectively, were used in the actual physical implementation. A signal called flip was used to switch from time1 to time2, and vice-versa. Notice that the corresponding decimals are marked beside each value of dout, so they can be easily verified in the simulation results.

```
1   ------------------------------------------------------------
2   LIBRARY ieee;
3   USE ieee.std_logic_1164.all;
4   ------------------------------------------------------------
5   ENTITY ssd_game2 IS
6      PORT ( clk, stop: IN BIT;
```

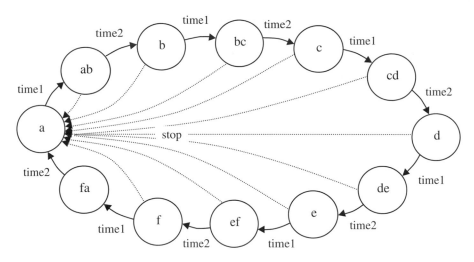

Figure 9.18
States diagram for the circuit of figure 9.17.

```
7                  dout: OUT BIT_VECTOR (6 DOWNTO 0));
8  END ssd_game2;
9  --------------------------------------------------------
10 ARCHITECTURE fsm OF ssd_game2 IS
11    CONSTANT time1: INTEGER := 4;   -- actual value is 80
12    CONSTANT time2: INTEGER := 2;   -- actual value is 30
13    TYPE states IS (a, ab, b, bc, c, cd, d, de, e, ef, f, fa);
14    SIGNAL present_state, next_state: STATES;
15    SIGNAL count: INTEGER RANGE 0 TO 5;
16    SIGNAL flip: BIT;
17 BEGIN
18    ------- Lower section of FSM (Sec. 8.2): ------------
19    PROCESS (clk, stop)
20    BEGIN
21       IF (stop='1') THEN
22          present_state <= a;
23       ELSIF (clk'EVENT AND clk='1') THEN
24          IF ((flip='1' AND count=time1) OR
25             (flip='0' AND count=time2)) THEN
26             count <= 0;
```

```
27              present_state <= next_state;
28          ELSE count <= count + 1;
29          END IF;
30      END IF;
31   END PROCESS;
32   ------- Upper section of FSM (Sec. 8.2): ------------
33   PROCESS (present_state)
34   BEGIN
35      CASE present_state IS
36          WHEN a =>
37              dout <= "1000000"; -- Decimal 64
38              flip<='1';
39              next_state <= ab;
40          WHEN ab =>
41              dout <= "1100000"; -- Decimal 96
42              flip<='0';
43              next_state <= b;
44          WIIEN b =>
45              dout <= "0100000"; -- Decimal 32
46              flip<='1';
47              next_state <= bc;
48          WHEN bc =>
49              dout <= "0110000"; -- Decimal 48
50              flip<='0';
51              next_state <= c;
52          WHEN c =>
53              dout <= "0010000"; -- Decimal 16
54              flip<='1';
55              next_state <= cd;
56          WHEN cd =>
57              dout <= "0011000"; -- Decimal 24
58              flip<='0';
59              next_state <= d;
60          WHEN d =>
61              dout <= "0001000"; -- Decimal 8
62              flip<='1';
63              next_state <= de;
```

```
64                WHEN de =>
65                    dout <= "0001100"; -- Decimal 12
66                    flip<='0';
67                    next_state <= e;
68                WHEN e =>
69                    dout <= "0000100"; -- Decimal 4
70                    flip<='1';
71                    next_state <= ef;
72                WHEN ef =>
73                    dout <= "0000110"; -- Decimal 6
74                    flip<='0';
75                    next_state <= f;
76                WHEN f =>
77                    dout <= "0000010"; -- Decimal 2
78                    flip<='1';
79                    next_state <= fa;
80                WHEN fa =>
81                    dout <= "1000010"; -- Decimal 66
82                    flip<='0';
83                    next_state <= a;
84            END CASE;
85        END PROCESS;
86 END fsm;
87 ----------------------------------------------------------
```

Simulation results are presented in figure 9.19. As can be seen, the system stays in the single states, a, b, etc., for four clock cycles (time1 = 4 here) and in the combined states, ab, bc, etc., for two clock cycles (time2 = 2). Observe also that the decimals detected by the simulator match the decimals listed in the VHDL code.

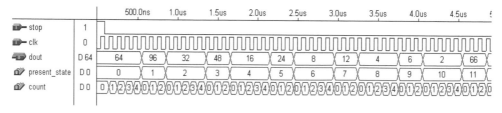

Figure 9.19
Simulation results of little SSD game of figure 9.17.

9.9 Signal Generators

Say that, from a clock signal (clk), we want to obtain the waveform shown in figure 9.20. In this kind of problem, we can use either the FSM approach or a conventional approach. Both kinds of solutions are illustrated below.

FSM Approach

The signal of figure 9.20 can be modeled as an 8-state FSM. Using a counter from 0 to 7, we can establish that wave = '0' (1st pulse) when count = 0, wave = '1' (2nd pulse) when count = 1, and so on, thus creating the signal shown in the figure. This implementation requires a total of four flip-flops: three to store count (three bits), plus one to store wave (one bit). Recall from chapter 8, sections 8.2–8.3, that the output of a FSM will only be registered if design style #2 is employed, which is necessary here, because glitches are not acceptable in a signal generator.

The corresponding VHDL code, using dsign style #2 (section 8.3), is shown below. Simulation results appear in figure 9.21. Checking the report file created by the synthesis tool, we verify that a total of four flip-flops were indeed inferred from this code.

```
1    ------------------------------------------------
2    LIBRARY ieee;
3    USE ieee.std_logic_1164.all;
```

Figure 9.20
Signal generator problem.

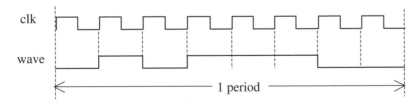

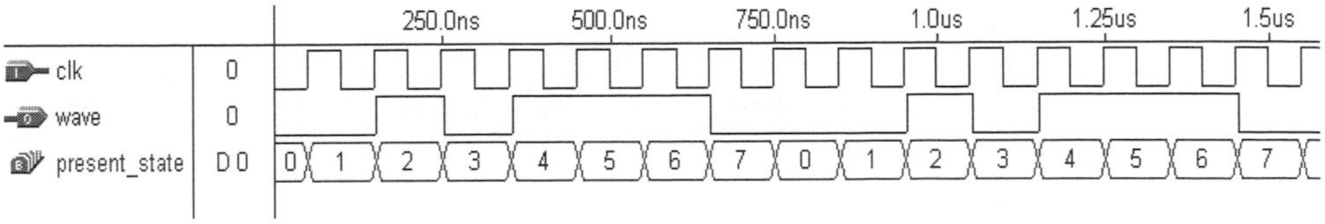

Figure 9.21
Simulation results of signal generator (FSM approach).

```
4  --------------------------------------------------------
5  ENTITY signal_gen IS
6     PORT (clk: IN STD_LOGIC;
7            wave: OUT STD_LOGIC);
8  END signal_gen;
9  --------------------------------------------------------
10 ARCHITECTURE fsm OF signal_gen IS
11    TYPE states IS (zero, one, two, three, four, five, six,
12                      seven);
13    SIGNAL present_state, next_state: STATES;
14    SIGNAL temp: STD_LOGIC;
15 BEGIN
16
17    --- Lower section of FSM (Sec. 8.3): ---
18    PROCESS (clk)
19    BEGIN
20       IF (clk'EVENT AND clk='1') THEN
21          present_state <= next_state;
22          wave <= temp;
23       END IF;
24    END PROCESS;
25
26    --- Upper section of FSM (Sec. 8.3): ---
27    PROCESS (present_state)
28    BEGIN
29       CASE present_state IS
30          WHEN zero => temp<='0'; next_state <= one;
31          WHEN one => temp<='1'; next_state <= two;
32          WHEN two => temp<='0'; next_state <= three;
33          WHEN three => temp<='1'; next_state <= four;
34          WHEN four => temp<='1'; next_state <= five;
35          WHEN five => temp<='1'; next_state <= six;
36          WHEN six => temp<='0'; next_state <= seven;
37          WHEN seven => temp<='0'; next_state <= zero;
38       END CASE;
39    END PROCESS;
40 END fsm;
41 --------------------------------------------------------
```

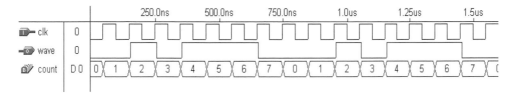

Figure 9.22
Simulation results of signal generator (conventional approach).

Conventional Approach

A conventional design, with the IF statement, is shown next. Notice that count and wave are both assigned at the transition of another signal (clk). Therefore, according to what you saw in section 7.5, both will be stored (that is, four flip-flops will be inferred, three for count and one for wave). Simulation results are shown in figure 9.22.

```
1    ------------------------------------
2    LIBRARY ieee;
3    USE ieee.std_logic_1164.all;
4    ------------------------------------
5    ENTITY signal_gen1 IS
6       PORT (clk: IN BIT;
7             wave: OUT BIT);
8    END signal_gen1;
9    ------------------------------------
10   ARCHITECTURE arch1 OF signal_gen1 IS
11   BEGIN
12      PROCESS
13         VARIABLE count: INTEGER RANGE 0 TO 7;
14      BEGIN
15         WAIT UNTIL (clk'EVENT AND clk='1');
16         CASE count IS
17            WHEN 0 => wave <= '0';
18            WHEN 1 => wave <= '1';
19            WHEN 2 => wave <= '0';
20            WHEN 3 => wave <= '1';
21            WHEN 4 => wave <= '1';
22            WHEN 5 => wave <= '1';
23            WHEN 6 => wave <= '0';
```

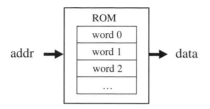

Figure 9.23
ROM diagram.

```
24               WHEN 7 => wave <= '0';
25          END CASE;
26          count := count + 1;
27      END PROCESS;
28 END arch1;
29 -------------------------------------
```

9.10 Memory Design

In this section, the design of the following memory circuits is presented:

· ROM
· RAM with separate in/out data buses
· RAM with bidirectional in/out data bus

ROM (Read Only Memory)

Figure 9.23 shows the diagram of a ROM. Since it is a read-only memory, no clock signal or write-enable pin is necessary. As can be seen, the circuit contains a pile of pre-stored words, being the one selected by the address input (addr) presented at the output (data).

In the code shown below, *words* (line 7) represents the number of words stored in the memory, while *bits* (line 6) represents the size of each word. To create a ROM, an array of CONSTANT values can be used (lines 15–22). First, a new TYPE, called *vector_array*, was defined (lines 13–14), which was then used in the declaration of a CONSTANT named *memory* (line 15). An 8×8 ROM is illustrated in this example, with the following (decimal) values stored in addresses 0 to 7: 0, 2, 4, 8, 16, 32, 64, and 128 (lines 15–22). Line 24 shows an example of call to the memory; the output (data) is equal to the word stored at address addr. When implementing a ROM, no registers are inferred, because no signal assignment occurs at the transition of another signal. Logical gates, forming an LUT (lookup table), are used instead.

```
1  ---------------------------------------------------
2  LIBRARY ieee;
3  USE ieee.std_logic_1164.all;
4  ---------------------------------------------------
5  ENTITY rom IS
6     GENERIC ( bits: INTEGER := 8;    -- # of bits per word
7               words: INTEGER := 8);  -- # of words in the memory
8     PORT ( addr: IN INTEGER RANGE 0 TO words-1;
9            data: OUT STD_LOGIC_VECTOR (bits-1 DOWNTO 0));
10 END rom;
11 ---------------------------------------------------
12 ARCHITECTURE rom OF rom IS
13    TYPE vector_array IS ARRAY (0 TO words-1) OF
14       STD_LOGIC_VECTOR (bits-1 DOWNTO 0);
15    CONSTANT memory: vector_array := (  "00000000",
16                                        "00000010",
17                                        "00000100",
18                                        "00001000",
19                                        "00010000",
20                                        "00100000",
21                                        "01000000",
22                                        "10000000");
23 BEGIN
24    data <= memory(addr);
25 END rom;
26 ---------------------------------------------------
```

Simulation results are shown in figure 9.24. As can be seen, the address changes from 0 to 7, then restarts from 0, with the outputs matching the values listed in the code above.

RAM with Separate Input and Output Data Buses

A RAM (Random Access Memory), with separate input and output data buses, is illustrated in figure 9.25. Indeed, this circuit was already discussed in example 6.11, but was repeated here to ease the comparison with the other memory circuits presented in this section.

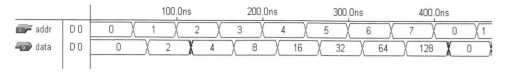

Figure 9.24
Simulation results from the 8 × 8 ROM code shown above.

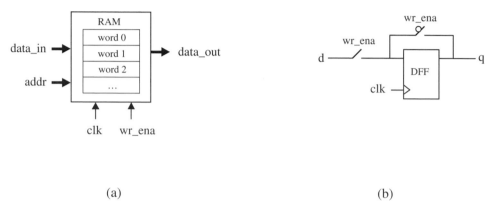

(a)   (b)

Figure 9.25
RAM with separate in/out data buses.

   As can be seen in figure 9.25(a), the circuit has a data input bus (data_in), a data output bus (data_out), an address bus (addr), plus clock (clk) and write enable (wr_ena) pins. When wr_enable is asserted, at the next rising edge of clk the vector present at data_in must be stored in the position specified by addr. data_out, on the other hand, must constantly display the data selected by addr.

   From the register point-of-view, the circuit can be summarized as in figure 9.25(b). When wr_ena is low, q is connected to the input of the flip-flop, and terminal d is open, so no new data will be written into the memory. However, when wr_ena is turned high, d is connected to the input of the register, so at the next rising edge of clk d will be stored.

   A VHDL code that implements the circuit of figure 9.25 is shown below. The chosen capacity was 16 words of length eight bits each. Notice that the code is totally generic. Simulation results are shown in figure 9.26.

```
1    ------------------------------------------------------
2    LIBRARY ieee;
```

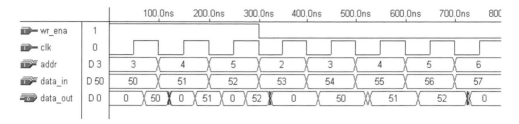

Figure 9.26
Simulation results of 16 × 8 RAM with separate in/out data buses.

```
3   USE ieee.std_logic_1164.all;
4   -----------------------------------------------------
5   ENTITY ram IS
6      GENERIC ( bits: INTEGER := 8;      -- # of bits per word
7                words: INTEGER := 16);   -- # of words in the
8                                         -- memory
9      PORT ( wr_ena, clk: IN STD_LOGIC;
10            addr: IN INTEGER RANGE 0 TO words-1;
11            data_in: IN STD_LOGIC_VECTOR (bits-1 DOWNTO 0);
12            data_out: OUT STD_LOGIC_VECTOR (bits-1 DOWNTO 0));
13  END ram;
14  -----------------------------------------------------
15  ARCHITECTURE ram OF ram IS
16     TYPE vector_array IS ARRAY (0 TO words-1) OF
17        STD_LOGIC_VECTOR (bits-1 DOWNTO 0);
18     SIGNAL memory: vector_array;
19  BEGIN
20     PROCESS (clk, wr_ena)
21     BEGIN
22        IF (wr_ena='1') THEN
23           IF (clk'EVENT AND clk='1') THEN
24              memory(addr) <= data_in;
25           END IF;
26        END IF;
27     END PROCESS;
28     data_out <= memory(addr);
29  END ram;
30  -----------------------------------------------------
```

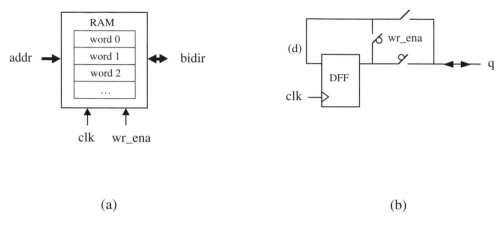

Figure 9.27
RAM with bidirectional in/out data bus.

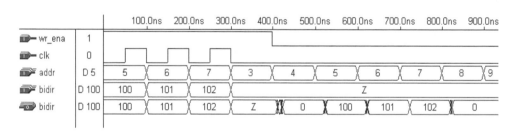

Figure 9.28
Simulation results of 16×8 RAM with bidirectional in/out data bus.

RAM with Bidirectional In/Out Data Bus

A RAM with bidirectional in/out data bus is illustrated in figure 9.27. The overall structure is similar to that of figure 9.25, except for the fact that now the same bus (bidir) is used to write data into the memory as well to read data from it.

From the register point-of-view, the circuit can be summarized as in figure 9.27(b). When wr_ena is low, the output of the register is connected to its input, so no change on the store data will occur. On the other hand, when wr_ena is asserted, q is connected to d, allowing new data to be stored at the next rising edge of clk.

A VHDL code that implements the circuit of figure 9.27 is shown below. The chosen capacity was 16 words of length eight bits each. Notice that this code is also totally generic. Simulation results are shown in figure 9.28.

```
 1  ---------------------------------------------
 2  LIBRARY ieee;
 3  USE ieee.std_logic_1164.all;
 4  ---------------------------------------------
 5  ENTITY ram4 IS
 6     GENERIC ( bits: INTEGER := 8;      -- # of bits per word
 7                words: INTEGER := 16);  -- # of words in the
 8                                        -- memory
 9     PORT ( clk, wr_ena: IN STD_LOGIC;
10            addr: IN INTEGER RANGE 0 TO words-1;
11            bidir: INOUT STD_LOGIC_VECTOR (bits-1 DOWNTO 0));
12  END ram4;
13  ---------------------------------------------
14  ARCHITECTURE ram OF ram4 IS
15     TYPE vector_array IS ARRAY (0 TO words-1) OF
16        STD_LOGIC_VECTOR (bits-1 DOWNTO 0);
17     SIGNAL memory: vector_array;
18  BEGIN
19     PROCESS (clk, wr_ena)
20     BEGIN
21        IF (wr_ena='0') THEN
22           bidir <= memory(addr);
23        ELSE
24           bidir <= (OTHERS => 'Z');
25           IF (clk'EVENT AND clk='1') THEN
26              memory(addr) <= bidir;
27           END IF;
28        END IF;
29     END PROCESS;
30  END ram;
31  ---------------------------------------------
```

9.11 Problems

Problem 9.1: Barrel Shifter

Why can we not replace the ARCHITECTURE of the barrel shifter presented in section 9.1 by that shown below, which is much shorter?

```
---------------------------------------------
ARCHITECTURE barrel OF barrel IS
BEGIN
   PROCESS (inp, shift)
   BEGIN
      IF (shift=0) THEN
         outp <= inp;
      ELSE
         FOR i IN 0 TO shift-1 LOOP
            outp(i) <= '0';
         END LOOP;
         FOR i IN shift TO inp'HIGH LOOP
            outp(i) <= inp(i-1);
         END LOOP;
      END IF;
   END PROCESS;
END barrel;
---------------------------------------------
```

Problem 9.2:   Divider

In section 9.4, we studied the design of fixed-point dividers. Two solutions were presented, both using sequential statements (IF and LOOP). Moreover, the codes implemented the second description of the division algorithm presented in that section. You are asked to write a *concurrent* solution for the division problem (with GENERATE). Additionally, your code should resemble the *first* description of the division algorithm (figure 9.9). In order to do so, we suggest the creation and use of the following types and signal:

```
SUBTYPE long IS STD_LOGIC_VECTOR (2n DOWNTO 0);
TYPE vec_array IS ARRAY (n DOWNTO 0) OF long;
SIGNAL a_input, b_input: vec_array;
```

where n should be declared as a GENERIC parameter.

Problem 9.3:   Vending-Machine Controller

Consider the vending-machine controller designed in section 9.5. We want to introduce some sophistications in it.

(a) In order to provide the necessary security, introduce some kind of handshake between the controller and the external circuitry. As an example, the handshake could include the following:

(i) an "input valid" signal (call it coin_valid), from the external circuit to the controller, informing that a new input is ready to be read. This signal should return to '0' as soon as it has been processed by the controller, so a new input will only be considered by the controller at its rising edge. This is important to avoid possible confusion which may occur when nickel_in, dime_in, or quarter_in stays present at the input of the FSM for more than one clock cycle (so it will not be interpreted as a second coin, as in the design of section 9.5)

(ii) an "input accepted" signal (call it coin_accepted), from the controller to the external circuit, informing that the present input has already been processed. Upon receiving this signal, the external circuit should cause coin_valid to return to '0'.

(b) Consider that the nickel or the dime box in the vending machine might run out of coins. Design alternative return paths taking such possibilities into consideration. (Suggestion: simply include new arrows between st45 → st40 and st35 → st30 in the FSM diagram of figure 9.11).

(c) Finally, consider the situation where a customer might continue depositing coins even when the necessary amount has already been reached. What should be done in such a situation?

Problem 9.4:   Serial Data Receiver

Try to model and design the serial data receiver of section 9.6 utilizing the FSM (finite state machine) approach (chapter 8). Before you start writing you VHDL code, present a clear states diagram of the system.

Problem 9.5:   Serial Data Transmitter

This problem is the counterpart of that treated in section 9.6. Here, the stored data must be *transmitted* serially. A diagram of the circuit is shown in figure P9.5. The

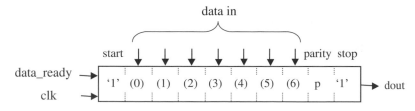

Figure P9.5

protocol is the same 10-bit structure of section 9.6; that is, a start bit (high), followed by seven bits of actual data, plus a parity bit, computed such that the total number of '1's in positions 2 to 9 is even, and finally a stop bit (also '1'). Consider that a data_ready signal is available to inform when the data can be loaded into the registers and sent out.

Problem 9.6:   Playing with an SSD

You are asked to introduce additional features in the little seven-segment display (SSD) game of section 9.8.

(a) Add a 2-bit input, called "speed", which should be able to select four different speeds for the circulatory movement. Keep the overlap time (time2) fixed at 30 ms, changing only time1. Choose four different circulatory periods and physically verify whether the circuit behaves as expected.

(b) Change the functionality of the stop input, such that instead of going to state when a stop is asserted, it freezes in whatever state it was when stop was activated, proceeding from there when stop returns to zero.

(c) Finally, add a "direction" pin. When low, the circuit should behave as above, but when high, it should circulate in the opposite direction (counterclockwise).

(d) Physical verification: After synthesizing and simulating your design, physically implement it in you PLD/FPGA development kit, following the steps below.

(i) First, verify in the report file generated by the compiler which pins of the chip were assigned to the inputs (clock and switches) and to the outputs (SSD).

(ii) Next, connect the signal generator (set to 1 kHz, with the appropriate logic levels, but leave it OFF while you make the connection) and the switches (which should provide VDD and GND levels) to the inputs of the circuit (your development kit board is normally equipped with test switches).

(iii) Connect the outputs of the chip to the SSD (your development kit board is normally equipped with seven-segment displays).

(iv) Finally, download the compiled file from your computer to the development kit, turn ON the signal generator, and verify the operation of your circuit. Play with the switches in order to test all operation modes.

Problem 9.7:   Speed Monitor

Figure P9.7 shows a possible view of a car speed monitor. The specifications of the system are the following:

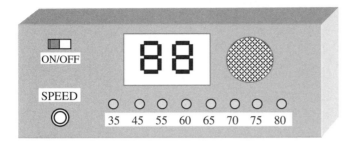

Figure P9.7

• Speed selection button (SPEED), which, at each touch, selects the next speed to be monitored (35, 45, 55, 60, 65, 70, 75, or 80 miles/hour).

• Set of eight LEDs, one for each speed. The LED corresponding to the selected speed should be ON.

• Two SSDs, which show the actual speed of the car. The car's electronic speedometer provides a clock signal whose frequency is proportional to the speed. You may check the data sheet of the speedometer that you are going to use, or you can start with a simple round number, which you can provide with a signal generator to test your circuit (say, 100 Hz per mile/hour).

• Buzzer, which emits alarm signals as the car approaches the selected speed. A 2 Hz signal should be emitted when the speed is three miles/hour or less from the selected speed, or a continuous alarm when at or above the selected speed. Consider a buzzer with internal oscillator, so only a DC signal must be provided in the latter case, or a square wave with frequency 2 Hz in the former case.

Write a VHDL code for such a circuit. Synthesize and simulate it. Finally, physically implement it in your PLD/FPGA development kit, using a signal generator for clock and following steps similar to those in problem 9.6.

Problem 9.8:   Random Number Generator

Design a 1-digit random number generator. The number should be from "0000" (display = 0) to "1111" (display = F). Use the circuit of section 9.8, with a modified function for the stop switch. The SSD should remain in a circular motion until the switch is pressed. When pressed, a random number should be displayed, being the circular movement resumed at the next touch of the switch. After compiling and simulating your circuit, physically implement it in your PDD/FPGA development kit.

II SYSTEM DESIGN

10 Packages and Components

10.1 Introduction

In Part I of the book, we studied the entire background and coding techniques of VHDL, which included the following:

- Code structure: library declarations, entity, architecture (chapter 2)
- Data types (chapter 3)
- Operators and attributes (chapter 4)
- Concurrent statements and concurrent code (chapter 5)
- Sequential statements and sequential code (chapter 6)
- Signals, variables, and constants (chapter 7)
- Design of finite state machines (chapter 8)
- Additional circuit designs (chapter 9)

Thus, in terms of figure 10.1, we may say that we have covered in detail all that is needed to construct the type of code depicted on its left-hand side. A good understanding of that material is indispensable, regardless of the design being just a small circuit or a very large system.

In Part II, we will simply add new building blocks to the material already presented. These new building blocks are intended mainly for library allocation, being shown on the right-hand side of figure 10.1. They are:

- Packages (chapter 10)
- Components (chapter 10)
- Functions (chapter 11)
- Procedures (chapter 11)

These new units can be located in the main code itself (that is, on the left-hand side of figure 10.1). However, since their main purpose is to allow common pieces of code to be reused and shared, it is more usual to place them in a LIBRARY. This also leads to code partitioning, which is helpful when dealing with long codes. In summary, frequently used pieces of code can be written in the form of COMPONENTS, FUNCTIONS, or PROCEDURES, then placed in a PACKAGE, which is finally compiled into the destination LIBRARY.

We have already seen (chapter 2) that at least three LIBRARIES are generally needed in a design: *ieee*, *std*, and *work*. After studying Part II, we will be able to construct our own libraries, which can then be added to the list above.

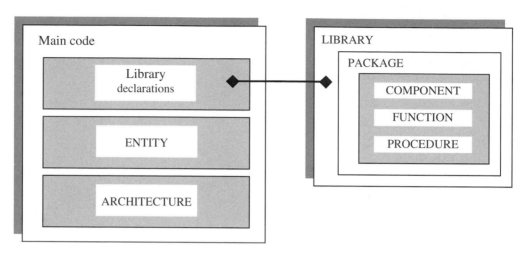

Figure 10.1
Fundamental units of VHDL code.

10.2   PACKAGE

As mentioned above, frequently used pieces of VHDL code are usually written in the form of COMPONENTS, FUNCTIONS, or PROCEDURES. Such codes are then placed inside a PACKAGE and compiled into the destination LIBRARY. The importance of this technique is that it allows code partitioning, code sharing, and code reuse.

We start by describing the structure of a PACKAGE. Besides COMPONENTS, FUNCTIONS, and PROCEDURES, it can also contain TYPE and CONSTANT definitions, among others. Its syntax is presented below.

```
PACKAGE package_name IS
    (declarations)
END package_name;

[PACKAGE BODY package_name IS
    (FUNCTION and PROCEDURE descriptions)
END package_name;]
```

As can be seen, the syntax is composed of two parts: PACKAGE and PACKAGE BODY. The first part is mandatory and contains all *declarations*, while the second part is necessary only when one or more subprograms (FUNCTION or PROCE-DURE) are declared in the upper part, in which case it must contain the *descriptions* (bodies) of the subprograms. PACKAGE and PACKAGE BODY must have the same name.

The declarations list can contain the following: COMPONENT, FUNCTION, PROCEDURE, TYPE, CONSTANT, etc.

Example 10.1:   Simple Package

The example below shows a PACKAGE called my_package. It contains only TYPE and CONSTANT declarations, so a PACKAGE BODY is not necessary.

```
1   ------------------------------------------------
2   LIBRARY ieee;
3   USE ieee.std_logic_1164.all;
4   ------------------------------------------------
5   PACKAGE my_package IS
6      TYPE state IS (st1, st2, st3, st4);
7      TYPE color IS (red, green, blue);
8      CONSTANT vec: STD_LOGIC_VECTOR(7 DOWNTO 0) := "11111111";
9   END my_package;
10  ------------------------------------------------
```

Example 10.2:   Package with a Function

This example contains, besides TYPE and CONSTANT declarations, a FUNC-TION. Therefore, a PACKAGE BODY is now needed (details on how to write a FUNCTION will be seen in chapter 11). This function returns TRUE when a positive edge occurs on clk.

```
1   ------------------------------------------------
2   LIBRARY ieee;
3   USE ieee.std_logic_1164.all;
4   ------------------------------------------------
5   PACKAGE my_package IS
6      TYPE state IS (st1, st2, st3, st4);
7      TYPE color IS (red, green, blue);
8      CONSTANT vec: STD_LOGIC_VECTOR(7 DOWNTO 0) := "11111111";
9      FUNCTION positive_edge(SIGNAL s: STD_LOGIC) RETURN BOOLEAN;
10  END my_package;
```

```
11 -------------------------------------------------
12 PACKAGE BODY my_package IS
13    FUNCTION positive_edge(SIGNAL s: STD_LOGIC) RETURN BOOLEAN IS
14    BEGIN
15       RETURN (s'EVENT AND s='1');
16    END positive_edge;
17 END my_package;
18 -------------------------------------------------
```

Any of the PACKAGES above (example 10.1 or example 10.2) can now be
compiled, becoming then part of our *work* LIBRARY (or any other). To make use
of it in a VHDL code, we have to add a new USE clause to the main code (USE
work.my_package.all), as shown below.

```
-----------------------------------
LIBRARY ieee;
USE ieee.std_logic_1164.all;
USE work.my_package.all;
-----------------------------------
ENTITY...
...
ARCHITECTURE...
...
-----------------------------------
```

10.3 COMPONENT

A COMPONENT is simply a piece of *conventional* code (that is, LIBRARY
declarations + ENTITY + ARCHITECTURE, as seen in chapter 2). However, by
declaring such code as being a COMPONENT, it can then be used within another
circuit, thus allowing the construction of *hierarchical* designs.

A COMPONENT is also another way of partitioning a code and providing code
sharing and code reuse. For example, commonly used circuits, like flip-flops, multiplexers, adders, basic gates, etc., can be placed in a LIBRARY, so any project can
make use of them without having to explicitly rewrite such codes.

To use (*instantiate*) a COMPONENT, it must first be *declared*. The corresponding
syntaxes are shown below.

COMPONENT declaration:

```
COMPONENT component_name IS
    PORT (
       port_name : signal_mode signal_type;
       port_name : signal_mode signal_type;
       ...);
END COMPONENT;
```

COMPONENT instantiation:

```
label: component_name PORT MAP (port_list);
```

As can be seen, the syntax of the declaration is similar to that of an ENTITY (section 2.3); that is, the names of the ports must be specified, along with their modes (IN, OUT, BUFFER, or INOUT) and data types (STD_LOGIC_VECTOR, INTEGER, BOOLEAN, etc.). To instantiate a component a label is required, followed by the component's name and a PORT MAP declaration. Finally, port_list is just a list relating the ports of the actual circuit to the ports of the pre-designed component which is being instantiated.

Example: Let us consider an inverter, which has been previously designed (inverter.vhd) and compiled into the work library. We can make use of it by means of the code shown below. The label chosen for this component was U1. The names of the ports in the actual circuit are x and y, which are being assigned to a and b, respectively, of the pre-designed inverter (this is called *positional* mapping, because the first signal in one corresponds to the first signal in the other, the second in one to the second in the other, and so on).

```
----- COMPONENT declaration: -----------
COMPONENT inverter IS
     PORT (a: IN STD_LOGIC; b: OUT STD_LOGIC);
END COMPONENT;

----- COMPONENT instantiation: -----------
U1: inverter PORT MAP (x, y);
```

There are two basic ways to declare a COMPONENT (figure 10.2). Once we have designed it and placed it in the destination LIBRARY, we can declare it in the main code itself, as shown in figure 10.2(a), or we can declare it using a PACKAGE, as in figure 10.2(b). The latter avoids the repetition of the declaration every time the COMPONENT is instantiated. Examples of both approaches are presented below.

Example 10.3:   Components Declared in the Main Code

We want to implement the circuit of figure 10.3 employing only COMPONENTS (inverter, nand_2, and nand_3), but without creating a specific PACKAGE to declare them, thus as in figure 10.2(a). Then four pieces of VHDL code are needed: one for each component, plus one for the project (main code). All four files are shown below. Notice that, since we have not created a PACKAGE, the COMPONENTS must be declared in the main code (in the declarative part of the ARCHITEC-TURE). Simulation results are presented in figure 10.4.

```
1   ------ File inverter.vhd: -------------------
2   LIBRARY ieee;
3   USE ieee.std_logic_1164.all;
4   -----------------------------------
5   ENTITY inverter IS
6     PORT (a: IN STD_LOGIC; b: OUT STD_LOGIC);
7   END inverter;
8   -----------------------------------
9   ARCHITECTURE inverter OF inverter IS
10  BEGIN
11    b <= NOT a;
12  END inverter;
13  -------------------------------------------

1   ------ File nand_2.vhd: --------------------
2   LIBRARY ieee;
3   USE ieee.std_logic_1164.all;
4   -----------------------------------
5   ENTITY nand_2 IS
6     PORT (a, b: IN STD_LOGIC; c: OUT STD_LOGIC);
7   END nand_2;
8   -----------------------------------
9   ARCHITECTURE nand_2 OF nand_2 IS
```

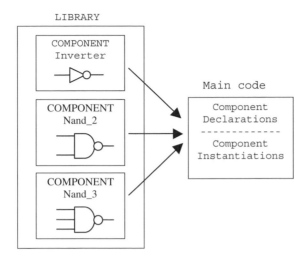

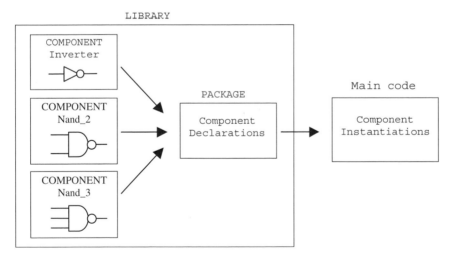

Figure 10.2
Basic ways of declaring COMPONENTS: (a) declarations in the main code itself, (b) declarations in a
PACKAGE.

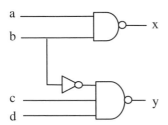

Figure 10.3
Circuit of example 10.3.

```
10 BEGIN
11    c <= NOT (a AND b);
12 END nand_2;
13 --------------------------------------------

1  ----- File nand_3.vhd: ----------------------
2  LIBRARY ieee;
3  USE ieee.std_logic_1164.all;
4  ------------------------------------
5  ENTITY nand_3 IS
6     PORT (a, b, c: IN STD_LOGIC; d: OUT STD_LOGIC);
7  END nand_3;
8  ------------------------------------
9  ARCHITECTURE nand_3 OF nand_3 IS
10 BEGIN
11    d <= NOT (a AND b AND c);
12 END nand_3;
13 --------------------------------------------

1  ----- File project.vhd: ---------------------
2  LIBRARY ieee;
3  USE ieee.std_logic_1164.all;
4  ------------------------------------
5  ENTITY project IS
6     PORT (a, b, c, d: IN STD_LOGIC;
7            x, y: OUT STD_LOGIC);
8  END project;
9  ------------------------------------
```

```
10 ARCHITECTURE structural OF project IS
11     -------------
12     COMPONENT inverter IS
13        PORT (a: IN STD_LOGIC; b: OUT STD_LOGIC);
14     END COMPONENT;
15     -------------
16     COMPONENT nand_2 IS
17        PORT (a, b: IN STD_LOGIC; c: OUT STD_LOGIC);
18     END COMPONENT;
19     -------------
20     COMPONENT nand_3 IS
21        PORT (a, b, c: IN STD_LOGIC; d: OUT STD_LOGIC);
22     END COMPONENT;
23     -------------
24     SIGNAL w: STD_LOGIC;
25 BEGIN
26     U1: inverter PORT MAP (b, w);
27     U2: nand_2 PORT MAP (a, b, x);
28     U3: nand_3 PORT MAP (w, c, d, y);
29 END structural;
30 --------------------------------------------
```

Example 10.4:   Components Declared in a Package

We want to implement the same project of the previous example (figure 10.3). However, we will now create a PACKAGE where all the COMPONENTS (inverter, nand_2, and nand_3) will be declared, like in figure 10.2(b). Thus now five pieces of VHDL code are needed: one for each component, one for the PACKAGE, and finally one for the project. Despite having an extra file (PACKAGE), such extra file needs to be created only once, thus avoiding the need to declare the components in the main code every time they are instantiated.

Notice that an extra USE clause (USE work.my_components.all) is now necessary, in order to make the PACKAGE my_components visible to the design. The simulation results are obviously the same as those of figure 10.4.

```
1  ------ File inverter.vhd: -------------------
2  LIBRARY ieee;
3  USE ieee.std_logic_1164.all;
4  ------------------------------------
```

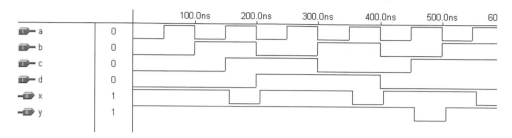

Figure 10.4
Experimental results of example 10.3.

```
5   ENTITY inverter IS
6      PORT (a: IN STD_LOGIC; b: OUT STD_LOGIC);
7   END inverter;
8   ------------------------------------
9   ARCHITECTURE inverter OF inverter IS
10  BEGIN
11     b <= NOT a;
12  END inverter;
13  ----------------------------------------------

1   ------ File nand_2.vhd: ---------------------
2   LIBRARY ieee;
3   USE ieee.std_logic_1164.all;
4   ------------------------------------
5   ENTITY nand_2 IS
6      PORT (a, b: IN STD_LOGIC; c: OUT STD_LOGIC);
7   END nand_2;
8   ------------------------------------
9   ARCHITECTURE nand_2 OF nand_2 IS
10  BEGIN
11     c <= NOT (a AND b);
12  END nand_2;
13  ----------------------------------------------

1   ----- File nand_3.vhd: ---------------------
2   LIBRARY ieee;
3   USE ieee.std_logic_1164.all;
4   ------------------------------------
```

```
5   ENTITY nand_3 IS
6      PORT (a, b, c: IN STD_LOGIC; d: OUT STD_LOGIC);
7   END nand_3;
8   ----------------------------------
9   ARCHITECTURE nand_3 OF nand_3 IS
10  BEGIN
11     d <= NOT (a AND b AND c);
12  END nand_3;
13  ------------------------------------------------

1   ----- File my_components.vhd: ---------------
2   LIBRARY ieee;
3   USE ieee.std_logic_1164.all;
4   -----------------------
5   PACKAGE my_components IS
6      ------ inverter: -------
7      COMPONENT inverter IS
8         PORT (a: IN STD_LOGIC; b: OUT STD_LOGIC);
9      END COMPONENT;
10     ------ 2-input nand: ---
11     COMPONENT nand_2 IS
12        PORT (a, b: IN STD_LOGIC; c: OUT STD_LOGIC);
13     END COMPONENT;
14     ------ 3-input nand: ---
15     COMPONENT nand_3 IS
16        PORT (a, b, c: IN STD_LOGIC; d: OUT STD_LOGIC);
17     END COMPONENT;
18     -----------------------
19  END my_components;
20  ----------------------------------------------

1   ----- File project.vhd: ---------------------
2   LIBRARY ieee;
3   USE ieee.std_logic_1164.all;
4   USE work.my_components.all;
5   ----------------------------------
6   ENTITY project IS
7      PORT ( a, b, c, d: IN STD_LOGIC;
8              x, y: OUT STD_LOGIC);
```

```
9   END project;
10  --------------------------------
11  ARCHITECTURE structural OF project IS
12     SIGNAL w: STD_LOGIC;
13  BEGIN
14     U1: inverter PORT MAP (b, w);
15     U2: nand_2 PORT MAP (a, b, x);
16     U3: nand_3 PORT MAP (w, c, d, y);
17  END structural;
18  --------------------------------------------
```

10.4 PORT MAP

There are two ways to map the PORTS of a COMPONENT during its instantiation: *positional* mapping and *nominal* mapping. Let us consider the following
example:

```
COMPONENT inverter IS
   PORT (a: IN STD_LOGIC; b: OUT STD_LOGIC);
END COMPONENT;
...
U1: inverter PORT MAP (x, y);
```

In it, the mapping is *positional*; that is, PORTS x and y correspond to a and b,
respectively. On the other hand, a *nominal* mapping would be the following:

```
U1: inverter PORT MAP (x=>a, y=>b);
```

Positional mapping is easier to write, but nominal mapping is less error-prone.
Ports can also be left unconnected (using the keyword OPEN). For example:

```
U2: my_circuit PORT MAP (x=>a, y=>b, w=>OPEN, z=>d);
```

10.5 GENERIC MAP

GENERIC units (discussed in section 4.5) can also be instantiated. In that case, a
GENERIC MAP must be used in the COMPONENT instantiation to pass information to the GENERIC parameters. The new syntax is shown below.

Figure 10.5
Generic parity generator to be instantiated in example 10.5.

```
label: compon_name GENERIC MAP (param. list) PORT MAP (port list);
```

As can be seen, the only differences from the syntax already presented are the inclusion of the word GENERIC and of a parameter list. The purpose is to inform that those parameters are to be considered as *generic*. The usage of GENERIC MAP is illustrated in the example below.

Example 10.5:   Instantiating a *Generic* Component

Let us consider the generic parity generator of example 4.3 (repeated in figure 10.5), which adds one bit to the input vector (on its left-hand side). Such bit must be a '0' if the number of '1's in the input vector is even, or a '1' if it is odd, such that the resulting vector will always contain an even number of '1's.

The code presented below is generic (that is, works for any positive integer n). Two files are shown: one relative to the COMPONENT (par_generator, which, indeed, we can assume as previously designed and available in the work library), and one relative to the project itself (main code), where the component par_generator is instantiated.

Notice that the default value (n = 7) of GENERIC in the COMPONENT file (parity_gen) will be overwritten by the value n = 2 passed to it by means of the GENERIC MAP statement in the COMPONENT instantiation. Notice also that the GENERIC declaration that appears along with the COMPONENT declaration in the second file is necessary, for it is part of the original (the component's) ENTITY. However, it is not necessary to declare its default value again. Simulation results from the circuit synthesized with the code below are shown in figure 10.6.

```
1    ------ File parity_gen.vhd (component): -------------
2    LIBRARY ieee;
3    USE ieee.std_logic_1164.all;
4    -----------------------------------
```

Figure 10.6
Simulation results of example 10.5.

```
5   ENTITY parity_gen IS
6      GENERIC (n : INTEGER := 7);   -- default is 7
7      PORT ( input: IN BIT_VECTOR (n DOWNTO 0);
8              output: OUT BIT_VECTOR (n+1 DOWNTO 0));
9   END parity_gen;
10  ----------------------------------
11  ARCHITECTURE parity OF parity_gen IS
12  BEGIN
13     PROCESS (input)
14        VARIABLE temp1: BIT;
15        VARIABLE temp2: BIT_VECTOR (output'RANGE);
16     BEGIN
17        temp1 := '0';
18        FOR i IN input'RANGE LOOP
19           temp1 := temp1 XOR input(i);
20           temp2(i) := input(i);
21        END LOOP;
22        temp2(output'HIGH) := temp1;
23        output <= temp2;
24     END PROCESS;
25  END parity;
26  -------------------------------------------------------

1   ------ File my_code.vhd (actual project): ------------
2   LIBRARY ieee;
3   USE ieee.std_logic_1164.all;
4   ----------------------------------
5   ENTITY my_code IS
6      GENERIC (n : POSITIVE := 2); -- 2 will overwrite 7
7      PORT ( inp: IN BIT_VECTOR (n DOWNTO 0);
```

```
8                  outp: OUT BIT_VECTOR (n+1 DOWNTO 0));
9  END my_code;
10 ----------------------------------
11 ARCHITECTURE my_arch OF my_code IS
12 -----------------------
13    COMPONENT parity_gen IS
14       GENERIC (n : POSITIVE);
15       PORT (input: IN BIT_VECTOR (n DOWNTO 0);
16             output: OUT BIT_VECTOR (n+1 DOWNTO 0));
17    END COMPONENT;
18 -----------------------
19 BEGIN
20    C1: parity_gen GENERIC MAP(n) PORT MAP(inp, outp);
21 END my_arch;
22 ----------------------------------------------------
```

Example 10.6:   ALU Made of COMPONENTS

In example 5.5, the design of an ALU (Arithmetic Logic Unit) was presented (diagram repeated in figure 10.7). In that example, the code was self-contained (that is, no external COMPONENT, FUNCTION, or PROCEDURE was called). In the present example, however, we will assume that our library contains the three components (logic_unit, arith_unit, and mux) with which the ALU can be constructed.

In the code shown below, besides the main code (alu.vhd), we have also included the design of the three components mentioned above. As can be seen, the COMPONENTS were declared in the main code itself. Simulation results are shown in figure 10.8, which are similar to those of example 5.5.

```
1  -------- COMPONENT arith_unit: --------------------
2  LIBRARY ieee;
3  USE ieee.std_logic_1164.all;
4  USE ieee.std_logic_unsigned.all;
5  ----------------------------------------
6  ENTITY arith_unit IS
7     PORT ( a, b: IN STD_LOGIC_VECTOR (7 DOWNTO 0);
8            sel: IN STD_LOGIC_VECTOR (2 DOWNTO 0);
9            cin: IN STD_LOGIC;
10           x: OUT STD_LOGIC_VECTOR (7 DOWNTO 0));
11 END arith_unit;
```

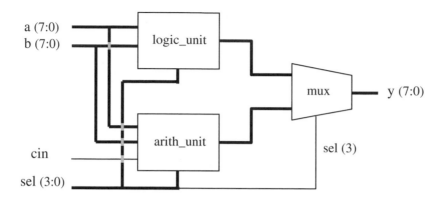

sel	Operation	Function	Unit
0000	y <= a	Transfer a	
0001	y <= a+1	Increment a	
0010	y <= a-1	Decrement a	
0011	y <= b	Transfer b	Arithmetic
0100	y <= b+1	Increment b	
0101	y <= b-1	Decrement b	
0110	y <= a+b	Add a and b	
0111	y <= a+b+cin	Add a and b with carry	
1000	y <= NOT a	Complement a	
1001	y <= NOT b	Complement b	
1010	y <= a AND b	AND	
1011	y <= a OR b	OR	Logic
1100	y <= a NAND b	NAND	
1101	y <= a NOR b	NOR	
1110	y <= a XOR b	XOR	
1111	y <= a XNOR b	XNOR	

Figure 10.7
ALU constructed from three COMPONENTS.

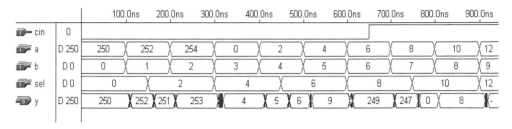

Figure 10.8
Simulation results of example 10.6.

```
12 -------------------------------------------
13 ARCHITECTURE arith_unit OF arith_unit IS
14    SIGNAL arith, logic: STD_LOGIC_VECTOR (7 DOWNTO 0);
15 BEGIN
16    WITH sel SELECT
17       x <=  a WHEN "000",
18             a+1 WHEN "001",
19             a-1 WHEN "010",
20             b WHEN "011",
21             b+1 WHEN "100",
22             b-1 WHEN "101",
23             a+b WHEN "110",
24             a+b+cin WHEN OTHERS;
25 END arith_unit;
26 -----------------------------------------------------

1  -------- COMPONENT logic_unit: --------------------
2  LIBRARY ieee;
3  USE ieee.std_logic_1164.all;
4  -------------------------------------------
5  ENTITY logic_unit IS
6     PORT ( a, b: IN STD_LOGIC_VECTOR (7 DOWNTO 0);
7            sel: IN STD_LOGIC_VECTOR (2 DOWNTO 0);
8            x: OUT STD_LOGIC_VECTOR (7 DOWNTO 0));
9  END logic_unit;
10 -------------------------------------------
11 ARCHITECTURE logic_unit OF logic_unit IS
12 BEGIN
```

```
13    WITH sel SELECT
14      x <=  NOT a WHEN "000",
15            NOT b WHEN "001",
16            a AND b WHEN "010",
17            a OR b WHEN "011",
18            a NAND b WHEN "100",
19            a NOR b WHEN "101",
20            a XOR b WHEN "110",
21            NOT (a XOR b) WHEN OTHERS;
22 END logic_unit;
23 ----------------------------------------------------

1  -------- COMPONENT mux: --------------------------
2  LIBRARY ieee;
3  USE ieee.std_logic_1164.all;
4  -----------------------------------------
5  ENTITY mux IS
6     PORT ( a, b: IN STD_LOGIC_VECTOR (7 DOWNTO 0);
7            sel: IN STD_LOGIC;
8            x: OUT STD_LOGIC_VECTOR (7 DOWNTO 0));
9  END mux;
10 -----------------------------------------
11 ARCHITECTURE mux OF mux IS
12 BEGIN
13   WITH sel SELECT
14      x <=   a WHEN '0',
15             b WHEN OTHERS;
16 END mux;
17 ----------------------------------------------------

1  -------- Project ALU (main code): -----------------
2  LIBRARY ieee;
3  USE ieee.std_logic_1164.all;
4  -----------------------------------------
5  ENTITY alu IS
6     PORT ( a, b: IN STD_LOGIC_VECTOR(7 DOWNTO 0);
7            cin: IN STD_LOGIC;
8            sel: IN STD_LOGIC_VECTOR(3 DOWNTO 0);
9            y: OUT STD_LOGIC_VECTOR(7 DOWNTO 0));
```

```
10 END alu;
11 -----------------------------------------
12 ARCHITECTURE alu OF alu IS
13 ----------------------
14    COMPONENT arith_unit IS
15    PORT ( a, b: IN STD_LOGIC_VECTOR(7 DOWNTO 0);
16           cin: IN STD_LOGIC;
17           sel: IN STD_LOGIC_VECTOR(2 DOWNTO 0);
18           x: OUT STD_LOGIC_VECTOR(7 DOWNTO 0));
19    END COMPONENT;
20    ----------------------
21    COMPONENT logic_unit IS
22    PORT ( a, b: IN STD_LOGIC_VECTOR(7 DOWNTO 0);
23           sel: IN STD_LOGIC_VECTOR(2 DOWNTO 0);
24           x: OUT STD_LOGIC_VECTOR(7 DOWNTO 0));
25    END COMPONENT;
26    ----------------------
27    COMPONENT mux IS
28    PORT ( a, b: IN STD_LOGIC_VECTOR(7 DOWNTO 0);
29           sel: IN STD_LOGIC;
30           x: OUT STD_LOGIC_VECTOR(7 DOWNTO 0));
31    END COMPONENT;
32    ----------------------
33    SIGNAL x1, x2: STD_LOGIC_VECTOR(7 DOWNTO 0);
34 ----------------------
35 BEGIN
36    U1: arith_unit PORT MAP (a, b, cin, sel(2 DOWNTO 0), x1);
37    U2: logic_unit PORT MAP (a, b, sel(2 DOWNTO 0), x2);
38    U3: mux PORT MAP (x1, x2, sel(3), y);
39 END alu;
40 -----------------------------------------------------
```

10.6   Problems

Problem 10.1:   ALU with Components Declared in a Package

Redo example 10.6. This time, create a PACKAGE containing all COMPONENT declarations. Then make the changes needed in the main code and recompile it. Synthesize and simulate your solution to fully verify its functionality.

Problem 10.2:   Carry-Ripple Adder Constructed from Components

Consider the carry-ripple adder discussed in section 9.3 (figure 9.6). Design a FAU (full-adder unit), to be used as a COMPONENT. Compile it into the *work* library. Then write a code for the complete carry-ripple adder containing instantiations of FAU. Compile your project and simulate the synthesized circuit, comparing the results with those obtained in section 9.3.

Problem 10.3:   Carry-Lookahead Adder Constructed from Components

Consider now the carry-lookahead adder of section 9.3 (figure 9.8). Design a PGU (propagate-generate unit) and a CLAU (carry-lookahead unit), to be used as COMPONENTS. Compile them into the *work* library. Then write a code for the complete carry-lookahead adder containing instantiations of PGU and CLAU. You can choose whether to declare the COMPONENTS in a specific PACKAGE or in the main code itself (in the declarative part of the ARCHITECTURE). Compile your project and simulate the synthesized circuit, comparing the results with those obtained in section 9.3.

Problem 10.4:   Registered Counter

Figure P10.4 illustrates the construction of a hierarchical design. Two sub-circuits (that is, "components"), called counter and register, are used to construct a higher-level circuit, called stop_watch. The system consists of a free-running counter, which is reset every time the stop input is asserted. The status of the counter must be stored in the sub-circuit register just before reset occurs. Once stop returns to '0', the counter resumes counting (from zero), while the register holds the previous count. Design the two components of figure P10.4, then instantiate them in the main code to produce the complete stop_watch circuit.

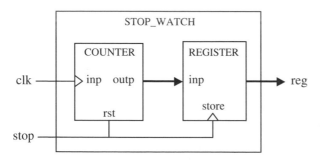

Figure P10.4

11 Functions and Procedures

FUNCTIONS and PROCEDURES are collectively called *subprograms*. From a construction point of view, they are very similar to a PROCESS (studied in chapter 6), for they are the only pieces of *sequential* VHDL code, and thus employ the same sequential statements seen there (IF, CASE, and LOOP; WAIT is not allowed). However, from the applications point of view, there is a fundamental difference between a PROCESS and a FUNCTION or PROCEDURE. While the first is intended for immediate use in the main code, the others are intended mainly for LIBRARY allocation, that is, their purpose is to store commonly used pieces of code, so they can be reused or shared by other projects. Nevertheless, if desired, a FUNCTION or PROCEDURE can also be installed in the main code itself.

11.1 FUNCTION

A FUNCTION is a section of *sequential* code. Its purpose is to create new functions to deal with commonly encountered problems, like data type conversions, logical operations, arithmetic computations, and new operators and attributes. By writing such code as a FUNCTION, it can be shared and reused, also propitiating the main code to be shorter and easier to understand.

As already mentioned, a FUNCTION is very similar to a PROCESS (section 6.1). The same statements that can be used in a process (IF, WAIT, CASE, and LOOP) can also be used in a function, with the exception of WAIT. Other two prohibitions in a function are SIGNAL declarations and COMPONENT instantiations.

To construct and use a function, two parts are necessary: the function itself (function body) and a call to the function. Their syntaxes are shown below.

Function Body

```
FUNCTION function_name [<parameter list>] RETURN data_type IS
    [declarations]
BEGIN
    (sequential statements)
END function_name;
```

In the syntax above, ⟨parameter list⟩ specifies the function's input parameters, that is:

⟨parameter list⟩ = [CONSTANT] constant_name: constant_type; or

⟨parameter list⟩ = SIGNAL signal_name: signal_type;

There can be any number of such parameters (even zero), which, as shown above, can only be CONSTANT (default) or SIGNAL (VARIABLES are not allowed). Their types can be any of the synthesizable data types studied in chapter 3 (BOO-LEAN, STD_LOGIC, INTEGER, etc.). However, no range specification should be included (for example, do not enter RANGE when using INTEGER, or TO/ DOWNTO when using STD_LOGIC_VECTOR). On the other hand, there is only one return value, whose type is specified by data_type.

Example: The function below, named f1, receives three parameters (a, b, and c). a and b are CONSTANTS (notice that the word CONSTANT can be omitted, for it is the default object), while c is a SIGNAL. a and b are of type INTEGER, while c is of type STD_LOGIC_VECTOR. Notice that neither RANGE nor DOWNTO was specified. The output parameter (there can be only one) is of type BOOLEAN.

```
FUNCTION f1 (a, b: INTEGER; SIGNAL c: STD_LOGIC_VECTOR)
    RETURN BOOLEAN IS
BEGIN
    (sequential statements)
END f1;
```

Function Call

A function is called as part of an expression. The expression can obviously appear by itself or associated to a statement (either concurrent or sequential).

Examples of function calls:

```
x <= conv_integer(a);        -- converts a to an integer
                             -- (expression appears by itself)
y <= maximum(a, b);          -- returns the largest of a and b
                             -- (expression appears by itself)
IF x > maximum(a, b) ...     -- compares x to the largest of a, b
                             -- (expression associated to a
                             -- statement)
```

Example 11.1:   Function positive_edge( )

The FUNCTION below detects a positive (rising) clock edge. It is similar to the IF(clk'EVENT and clk = '1') statement. This function could be used, for example, in the implementation of a DFF.

```
------ Function body: ------------------------------
FUNCTION positive_edge(SIGNAL s: STD_LOGIC) RETURN BOOLEAN IS
BEGIN
    RETURN (s'EVENT AND s='1');
END positive_edge;
------ Function call: ------------------------------
...
IF positive_edge(clk) THEN...
...
----------------------------------------------------
```

Example 11.2:   Function conv_integer( )

The FUNCTION presented next converts a parameter of type STD_LOGIC_
VECTOR into an INTEGER. Notice that the code is generic, that is, it works for
any range or order (TO/DOWNTO) of the input STD_LOGIC_VECTOR parameter. A typical call to the function is also shown.

```
------ Function body: ------------------------------
FUNCTION conv_integer (SIGNAL vector: STD_LOGIC_VECTOR)
      RETURN INTEGER IS
    VARIABLE result: INTEGER RANGE 0 TO 2**vector'LENGTH-1;
BEGIN
    IF (vector(vector'HIGH)='1') THEN result:=1;
    ELSE result:=0;
    END IF;
    FOR i IN (vector'HIGH-1) DOWNTO (vector'LOW) LOOP
       result:=result*2;
       IF(vector(i)='1') THEN result:=result+1;
       END IF;
    END LOOP;
    RETURN result;
END conv_integer;

------ Function call: ------------------------------
...
y <= conv_integer(a);
...
----------------------------------------------------
```

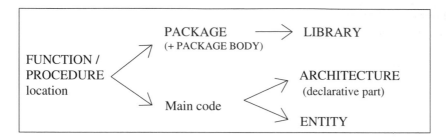

Figure 11.1
Typical locations of a FUNCTION or PROCEDURE.

11.2   Function Location

The typical locations of a FUNCTION (or PROCEDURE) are depicted in figure
11.1. Though a FUNCTION is usually placed in a PACKAGE (for code partitioning, code reuse, and code sharing purposes), it can also be located in the main code
(either inside the ARCHITECTURE or inside the ENTITY).

   When placed in a PACKAGE, then a PACKAGE BODY is necessary, which
must contain the body of each FUNCTION (or PROCEDURE) declared in the declarative part of the PACKAGE. Examples of both cases are presented below.

Example 11.3:   FUNCTION Located in the Main Code

Let us consider the positive_edge( ) function of example 11.1 As mentioned above,
when installed in the main code itself, the function can be located either in the
ENTITY or in the declarative part of the ARCHITECTURE. In the present example, the function appears in the latter, and is used to construct a DFF.

```
1   ----------------------------------------------
2   LIBRARY ieee;
3   USE ieee.std_logic_1164.all;
4   ----------------------------------------------
5   ENTITY dff IS
6      PORT ( d, clk, rst: IN STD_LOGIC;
7               q: OUT STD_LOGIC);
8   END dff;
9   ----------------------------------------------
10  ARCHITECTURE my_arch OF dff IS
11  ----------------------------------------------
```

```
12    FUNCTION positive_edge(SIGNAL s: STD_LOGIC)
13       RETURN BOOLEAN IS
14    BEGIN
15       RETURN s'EVENT AND s='1';
16    END positive_edge;
17 ---------------------------------------
18 BEGIN
19    PROCESS (clk, rst)
20    BEGIN
21       IF (rst='1') THEN q <= '0';
22       ELSIF positive_edge(clk) THEN q <= d;
23       END IF;
24    END PROCESS;
25 END my_arch;
26 -------------------------------------------
```

Example 11.4:   FUNCTION Located in a PACKAGE

This example is similar to example 11.3, with the only difference being that the FUNCTION located in a PACKAGE can now be reused and shared by other projects. Notice that, when placed in a PACKAGE, the function is indeed *declared* in the PACKAGE, but *described* in the PACKAGE BODY.

Below two VHDL codes are presented, being one relative to the construction of the FUNCTION / PACKAGE, while the other is an example where a call to the FUNCTION is made. The two codes can be compiled as two separate files, or can be compiled as a single file (saved as dff.vhd, which is the ENTITY's name). Notice the inclusion of "USE work.my_package.all;" in the main code (line 4).

```
1  ------- Package: ----------------------------
2  LIBRARY ieee;
3  USE ieee.std_logic_1164.all;
4  -------------------------------------------
5  PACKAGE my_package IS
6     FUNCTION positive_edge(SIGNAL s: STD_LOGIC) RETURN BOOLEAN;
7  END my_package;
8  -------------------------------------------
9  PACKAGE BODY my_package IS
10    FUNCTION positive_edge(SIGNAL s: STD_LOGIC)
11       RETURN BOOLEAN IS
```

```
12     BEGIN
13         RETURN s'EVENT AND s='1';
14     END positive_edge;
15 END my_package;
16 -----------------------------------------------

1  ------ Main code: --------------------------
2  LIBRARY ieee;
3  USE ieee.std_logic_1164.all;
4  USE work.my_package.all;
5  -----------------------------------------------
6  ENTITY dff IS
7     PORT ( d, clk, rst: IN STD_LOGIC;
8             q: OUT STD_LOGIC);
9  END dff;
10 -----------------------------------------------
11 ARCHITECTURE my_arch OF dff IS
12 BEGIN
13    PROCESS (clk, rst)
14    BEGIN
15       IF (rst='1') THEN q <= '0';
16       ELSIF positive_edge(clk) THEN  q <= d;
17       END IF;
18    END PROCESS;
19 END my_arch;
20 -----------------------------------------------
```

Example 11.5:   Function conv_integer( )

The conv_integer( ) function shown below was already seen in example 11.2; it converts a STD_LOGIC_VECTOR value into an INTEGER value. Below, the function was placed in a PACKAGE (plus PACKAGE BODY). A call to this function appears in the main code that follows the function implementation.

```
1  --------- Package: --------------------------
2  LIBRARY ieee;
3  USE ieee.std_logic_1164.all;
4  -----------------------------------------------
5  PACKAGE my_package IS
```

```
6     FUNCTION conv_integer (SIGNAL vector: STD_LOGIC_VECTOR)
7         RETURN INTEGER;
8  END my_package;
9  ------------------------------------------------
10 PACKAGE BODY my_package IS
11    FUNCTION conv_integer (SIGNAL vector: STD_LOGIC_VECTOR)
12           RETURN INTEGER IS
13       VARIABLE result: INTEGER RANGE 0 TO 2**vector'LENGTH-1;
14    BEGIN
15       IF (vector(vector'HIGH)='1') THEN result:=1;
16       ELSE result:=0;
17       END IF;
18       FOR i IN (vector'HIGH-1) DOWNTO (vector'LOW) LOOP
19           result:=result*2;
20           IF(vector(i)='1') THEN result:=result+1;
21           END IF;
22       END LOOP;
23       RETURN result;
24    END conv_integer;
25 END my_package;
26 ------------------------------------------------

1  -------- Main code: ------------------------
2  LIBRARY ieee;
3  USE ieee.std_logic_1164.all;
4  USE work.my_package.all;
5  ------------------------------------------------
6  ENTITY conv_int2 IS
7     PORT ( a: IN STD_LOGIC_VECTOR(0 TO 3);
8            y: OUT INTEGER RANGE 0 TO 15);
9  END conv_int2;
10 ------------------------------------------------
11 ARCHITECTURE my_arch OF conv_int2 IS
12 BEGIN
13    y <= conv_integer(a);
14 END my_arch;
15 ------------------------------------------------
```

Example 11.6:   Overloaded "+" Operator

The function shown below, called "+", overloads the pre-defined "+" (addition) operator (section 4.1 and section 4.4). Recall that the latter accepts only INTEGER, SIGNED, or UNSIGNED values. However, we are interested in writing a function which should allow the sum of STD_LOGIC_VECTOR values as well (thus over-loading the "+" operator).

The function shown below was placed in a PACKAGE (plus PACKAGE BODY). An example utilizing this function is also presented in the main code that follows the function implementation. Notice that the two parameters passed to the function, as well as the return value, are all of type STD_LOGIC_VECTOR. We assume that they all have the same number of bits (an extension to this example is presented in problem 11.8).

```
1  -------- Package: ----------------------------
2  LIBRARY ieee;
3  USE ieee.std_logic_1164.all;
4  ----------------------------------------------
5  PACKAGE my_package IS
6     FUNCTION "+" (a, b: STD_LOGIC_VECTOR)
7        RETURN STD_LOGIC_VECTOR;
8  END my_package;
9  ----------------------------------------------
10 PACKAGE BODY my_package IS
11    FUNCTION "+" (a, b: STD_LOGIC_VECTOR)
12          RETURN STD_LOGIC_VECTOR IS
13       VARIABLE result: STD_LOGIC_VECTOR;
14       VARIABLE carry: STD_LOGIC;
15    BEGIN
16       carry := '0';
17       FOR i IN a'REVERSE_RANGE LOOP
18          result(i) := a(i) XOR b(i) XOR carry;
19          carry := (a(i) AND b(i)) OR (a(i) AND carry) OR
20                   (b(i) AND carry);
21       END LOOP;
22       RETURN result;
23    END "+";
24 END my_package;
25 ----------------------------------------------
```

Figure 11.2
Simulation results of example 11.6.

```
1    --------- Main code: ------------------------
2    LIBRARY ieee;
3    USE ieee.std_logic_1164.all;
4    USE work.my_package.all;
5    ---------------------------------------------
6    ENTITY add_bit IS
7       PORT ( a: IN STD_LOGIC_VECTOR(3 DOWNTO 0);
8                y: OUT STD_LOGIC_VECTOR(3 DOWNTO 0));
9    END add_bit;
10   ---------------------------------------------
11   ARCHITECTURE my_arch OF add_bit IS
12      CONSTANT b: STD_LOGIC_VECTOR(3 DOWNTO 0) := "0011";
13      CONSTANT c: STD_LOGIC_VECTOR(3 DOWNTO 0) := "0110";
14   BEGIN
15      y <= a + b + c;     -- overloaded "+" operator
16   END my_arch;
17   ---------------------------------------------
```

Simulation results, for 4-bit numbers, are presented in figure 11.2. We have entered $b = 3$ and $c = 6$ as two constants, which are added to the input signal a. The expected results is then $y = a + 9$.

Example 11.7:   Arithmetic Shift Function

The function shown below arithmetically shifts a STD_LOGIC_VECTOR value to the left. Two arguments are passed to the function: arg1 and arg2. The first is the vector to be shifted, while the second specifies the amount of shift. Notice that the function (lines 13–26) is totally generic; that is, it works for any size (number of bits) or order (TO/DOWNTO) of the input vector. In this example, the function was located in the main code instead of in a package.

```
1   ----------------------------------------
2   LIBRARY ieee;
3   USE ieee.std_logic_1164.all;
4   ----------------------------------------
5   ENTITY shift_left IS
6      GENERIC (size: INTEGER := 4);
7      PORT ( a: IN STD_LOGIC_VECTOR(size-1 DOWNTO 0);
8               x, y, z: OUT STD_LOGIC_VECTOR(size-1 DOWNTO 0));
9   END shift_left;
10  ----------------------------------------
11  ARCHITECTURE behavior OF shift_left IS
12  ----------------------------------------
13     FUNCTION slar (arg1: STD_LOGIC_VECTOR; arg2: NATURAL)
14          RETURN STD_LOGIC_VECTOR IS
15        VARIABLE input: STD_LOGIC_VECTOR(size-1 DOWNTO 0) := arg1;
16        CONSTANT size : INTEGER := arg1'LENGTH;
17        VARIABLE copy: STD_LOGIC_VECTOR(size-1 DOWNTO 0)
18           := (OTHERS => arg1(arg1'RIGHT));
19        VARIABLE result: STD_LOGIC_VECTOR(size-1 DOWNTO 0);
20     BEGIN
21        IF (arg2 >= size-1) THEN result := copy;
22        ELSE result := input(size-1-arg2 DOWNTO 1) &
23           copy(arg2 DOWNTO 0);
24        END IF;
25        RETURN result;
26     END slar;
27  ----------------------------------------
28  BEGIN
29     x <= slar(a, 0);
30     y <= slar(a, 1);
31     z <= slar(a, 2);
32  END behavior;
33  ----------------------------------------
```

Simulation results are shown in figure 11.3 (for y only). The upper set of curves corresponds to the a(size-1 DOWNTO 0) specification, as shown above in line 7 (that is, a(3) is the MSB), while the second set refers to the reverse order, that is, a(0 TO 3), in which case a(0) is the MSB.

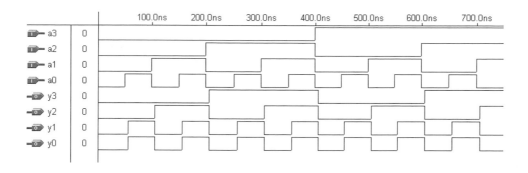

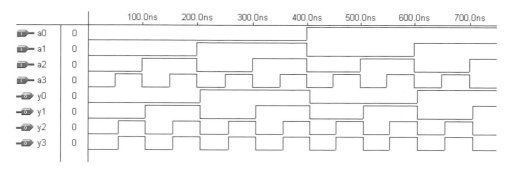

Figure 11.3
Simulation results of example 11.7.

Example 11.8:   Multiplier

In this example, a function called mult( ) is presented. It multiplies two UNSIGNED values, returning their UNSIGNED product. The parameters passed to the function do not need to have the same number of bits, and their order (TO/DOWNTO) can be any. The function was installed in a package called pack. An application example (main code) is also presented. Simulation results are shown in figure 11.4.

```
1    --------- Package: ---------------------------------
2    LIBRARY ieee;
3    USE ieee.std_logic_1164.all;
4    USE ieee.std_logic_arith.all;
5    ---------------------------------------------
6    PACKAGE pack IS
7       FUNCTION mult(a, b: UNSIGNED) RETURN UNSIGNED;
```

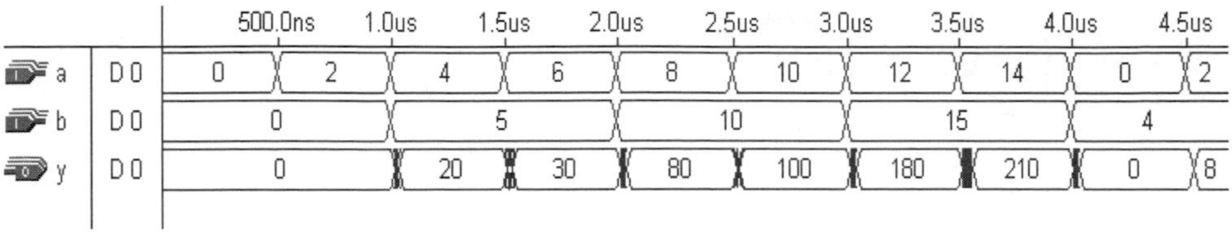

Figure 11.4
Simulation results of example 11.8.

```
8  END pack;
9  ----------------------------------------------
10 PACKAGE BODY pack IS
11    FUNCTION mult(a, b: UNSIGNED) RETURN UNSIGNED IS
12       CONSTANT max: INTEGER := a'LENGTH + b'LENGTH - 1;
13       VARIABLE aa: UNSIGNED(max DOWNTO 0) :=
14          (max DOWNTO a'LENGTH => '0')
15          & a(a'LENGTH-1 DOWNTO 0);
16       VARIABLE prod: UNSIGNED(max DOWNTO 0) := (OTHERS => '0');
17    BEGIN
18       FOR i IN 0 TO a'LENGTH-1 LOOP
19          IF (b(i)='1') THEN prod := prod + aa;
20          END IF;
21          aa := aa(max-1 DOWNTO 0) & '0';
22       END LOOP;
23       RETURN prod;
24    END mult;
25 END pack;
26 ----------------------------------------------------------

1  -------- Main code: ----------------------------------
2  LIBRARY ieee;
3  USE ieee.std_logic_1164.all;
4  USE ieee.std_logic_arith.all;
5  USE work.my_package.all;
6  ----------------------------------------------
7  ENTITY multiplier IS
8     GENERIC (size: INTEGER := 4);
9     PORT ( a, b: IN UNSIGNED(size-1 DOWNTO 0);
```

```
10               y: OUT UNSIGNED(2*size-1 DOWNTO 0));
11 END multiplier;
12 ------------------------------------------
13 ARCHITECTURE behavior OF multiplier IS
14 BEGIN
15    y <= mult(a,b);
16 END behavior;
17 ----------------------------------------------------
```

11.3 PROCEDURE

A PROCEDURE is very similar to a FUNCTION and has the same basic purposes. However, a procedure can return more than one value.

Like a FUNCTION, two parts are necessary to construct and use a PROCEDURE: the procedure itself (procedure body) and a procedure call.

Procedure Body

```
PROCEDURE procedure_name [<parameter list>] IS
    [declarations]
BEGIN
    (sequential statements)
END procedure_name;
```

In the syntax above, `<parameter list>` specifies the procedure's input and output parameters; that is:

⟨parameter list⟩ = [CONSTANT] constant_name: mode type;

⟨parameter list⟩ = SIGNAL signal_name: mode type; or

⟨parameter list⟩ = VARIABLE variable_name: mode type;

A PROCEDURE can have any number of IN, OUT, or INOUT parameters, which can be SIGNALS, VARIABLES, or CONSTANTS. For input signals (mode IN), the default is CONSTANT, whereas for output signals (mode OUT or INOUT) the default is VARIABLE.

As seen before, WAIT, SIGNAL declarations, and COMPONENTS are not synthesizable when used in a FUNCTION. The same is true for a PROCEDURE, with the exception that a SIGNAL can be declared, but then the PROCEDURE must be declared in a PROCESS. Moreover, besides WAIT, any other edge detection is also not synthesizable with a PROCEDURE (that is, contrary to a function, a synthesizable procedure should not infer registers).

In section 11.5, a summary comparing FUNCTIONS and PROCEDURES will be presented.

Example: The PROCEDURE below has three inputs, a, b, and c (mode IN). a is a CONSTANT of type BIT, while b and c are SIGNALS, also of type BIT. Notice that the word CONSTANT can be omitted for input parameters, for it is the default object (recall, however, that for outputs the default object is VARIABLE). There are also two return signals, x (mode OUT, type BIT_VECTOR) and y (mode INOUT, type INTEGER).

```
PROCEDURE my_procedure ( a: IN BIT; SIGNAL b, c: IN BIT;
                         SIGNAL x: OUT BIT_VECTOR(7 DOWNTO 0);
                         SIGNAL y: INOUT INTEGER RANGE 0 TO 99) IS
BEGIN
   ...
END my_procedure;
```

Procedure Call

Contrary to a FUNCTION, which is called as part of an expression, a PROCEDURE call is a statement on its own. It can appear by itself or associated to a statement (either concurrent or sequential).

Examples of procedure calls:

```
compute_min_max(in1, in2, 1n3, out1, out2);
            -- statement by itself

divide(dividend, divisor, quotient, remainder);
            -- statement by itself

IF (a>b) THEN compute_min_max(in1, in2, 1n3, out1, out2);
              -- procedure call associated to another statement
```

11.4   Procedure Location

The typical locations of a PROCEDURE are the same as those of a FUNCTION (see figure 11.1). Again, though it is usually placed in a PACKAGE (for code parti-

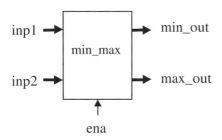

Figure 11.5
min_max circuit of example 11.9.

		100.0ns	200.0ns	300.0ns	400.0ns	500.0ns	600.0ns	700.0ns	800.0ns	900.0
inp1	D 0	0	20	40	60	80	100			
inp2	D 120	120	100	80	60	40	20			
ena	0									
min_out	D 0	0	20	40	60	40	20			
max_out	D 0	0	120	100	80	60	80	100		

Figure 11.6
Simulation results of example 11.9.

tioning, code reuse, and code sharing purposes), it can also be located in the main code (either in the ENTITY or in the declarative part of the ARCHITECTURE). When placed in a PACKAGE, a PACKAGE BODY is then necessary, which must contain the body of each PROCEDURE declared in the declarative part of the PACKAGE. Examples of both cases are shown below.

Example 11.9:   PROCEDURE Located in the Main Code

The min_max code below makes use of a PROCEDURE called sort. It takes two 8-bit unsigned integers as inputs (inp1, inp2), sorts them, then outputs the smaller value at min_out and the higher value at max_out (figure 11.5). The PROCEDURE is located in the declarative part of the ARCHITECTURE (main code). Notice that the PROCEDURE call, sort(inp1,inp2,min_out,max_out), is a statement on its own. Simulation results are shown in figure 11.6.

```
1   ------------------------------------------------------
2   LIBRARY ieee;
3   USE ieee.std_logic_1164.all;
```

```
4  -------------------------------------------------------
5  ENTITY min_max IS
6     GENERIC (limit : INTEGER := 255);
7     PORT ( ena: IN BIT;
8              inp1, inp2: IN INTEGER RANGE 0 TO limit;
9              min_out, max_out: OUT INTEGER RANGE 0 TO limit);
10 END min_max;
11 -------------------------------------------------------
12 ARCHITECTURE my_architecture OF min_max IS
13    -------------------------
14    PROCEDURE sort (SIGNAL in1, in2: IN INTEGER RANGE 0 TO limit;
15       SIGNAL min, max: OUT INTEGER RANGE 0 TO limit) IS
16    BEGIN
17       IF (in1 > in2) THEN
18          max <= in1;
19          min <= in2;
20       ELSE
21          max <= in2;
22          min <= in1;
23       END IF;
24    END sort;
25    -------------------------
26 BEGIN
27    PROCESS (ena)
28    BEGIN
29       IF (ena='1') THEN sort (inp1, inp2, min_out, max_out);
30       END IF;
31    END PROCESS;
32 END my_architecture;
33 -------------------------------------------------------
```

Example 11.10:   PROCEDURE Located in a PACKAGE

This example is similar to example 11.9, with the only difference being that now the PROCEDURE (called *sort*) is placed in a PACKAGE (called *my_package*). Thus the PROCEDURE can now be reused and shared with other designs. The code below can be compiled as two separate files, or can be compiled as a single file (called min_max.vhd, which is the ENTITY's name).

```
 1   ------------ Package: --------------------------
 2   LIBRARY ieee;
 3   USE ieee.std_logic_1164.all;
 4   -------------------------------------
 5   PACKAGE my_package IS
 6      CONSTANT limit: INTEGER := 255;
 7      PROCEDURE sort (SIGNAL in1, in2: IN INTEGER RANGE 0 TO limit;
 8         SIGNAL min, max: OUT INTEGER RANGE 0 TO limit);
 9   END my_package;
10   -------------------------------------
11   PACKAGE BODY my_package IS
12      PROCEDURE sort (SIGNAL in1, in2: IN INTEGER RANGE 0 TO limit;
13         SIGNAL min, max: OUT INTEGER RANGE 0 TO limit) IS
14      BEGIN
15         IF (in1 > in2) THEN
16            max <= in1;
17            min <= in2;
18         ELSE
19            max <= in2;
20            min <= in1;
21         END IF;
22      END sort;
23   END my_package;
24   --------------------------------------------------

 1   --------- Main code: ----------------------------
 2   LIBRARY ieee;
 3   USE ieee.std_logic_1164.all;
 4   USE work.my_package.all;
 5   ------------------------------------
 6   ENTITY min_max IS
 7      GENERIC (limit: INTEGER := 255);
 8      PORT ( ena: IN BIT;
 9             inp1, inp2: IN INTEGER RANGE 0 TO limit;
10             min_out, max_out: OUT INTEGER RANGE 0 TO limit);
11   END min_max;
12   -------------------------------------
```

```
13 ARCHITECTURE my_architecture OF min_max IS
14 BEGIN
15    PROCESS (ena)
16    BEGIN
17       IF (ena='1') THEN sort (inp1, inp2, min_out, max_out);
18       END IF;
19    END PROCESS;
20 END my_architecture;
21 ----------------------------------------------
```

The simulation results are obviously the same as those of example 11.9 (figure 11.6).

11.5 FUNCTION versus PROCEDURE Summary

• A FUNCTION has zero or more input parameters and a single return value. The input parameters can only be CONSTANTS (default) or SIGNALS (VARIABLES are not allowed).

• A PROCEDURE can have any number of IN, OUT, and INOUT parameters, which can be SIGNALS, VARIABLES, or CONSTANTS. For input parameters (mode IN) the default is CONSTANT, whereas for output parameters (mode OUT or INOUT) the default is VARIABLE.

• A FUNCTION is called as part of an expression, while a PROCEDURE is a statement on its own.

• In both, WAIT and COMPONENTS are not synthesizable.

• The possible locations of FUNCTIONS and PROCEDURES are the same (figure 11.1). Though they are usually placed in PACKAGES (for code partitioning, code sharing, and code reuse purposes), they can also be located in the main code (either inside the ARCHITECTURE or inside the ENTITY). When placed in a PACKAGE, then a PACKAGE BODY is necessary, which should contain the body of each FUNCTION and/or PROCEDURE declared in the PACKAGE.

11.6 ASSERT

ASSERT is a non-synthesizable statement whose purpose is to write out messages (on the screen, for example) when problems are found during simulation. Depending on the severity of the problem, the simulator is instructed to halt. Its syntax is the following:

```
ASSERT condition
[REPORT "message"]
[SEVERITY severity_level];
```

The severity level can be: Note, Warning, Error (default), or Failure. The message is written when the condition is FALSE.

Example: Say that we have written a function to add two binary numbers (like in example 11.6), where it was assumed that the input parameters must have the same number of bits. In order to check such an assumption, the following ASSERT statement could be included in the function body:

```
ASSERT a'LENGTH = b'LENGTH
REPORT "Error: vectors do not have same length!"
SEVERITY failure;
```

Again, ASSERT does not generate hardware. Synthesis tools will simply ignore it or give a warning.

11.7 Problems

The purpose of the problems proposed in this section is to reinforce the main aspects related to the construction and use of subprograms (FUNCTIONS and PROCEDURES).

Problem 11.1: Conversion to std_logic_vector

Write a function capable of converting an INTEGER to a STD_LOGIC_VECTOR value. Call it conv_std_logic( ). Then write an application example, containing a call to your function, in order to test it. Construct two solutions: one with the function in the main code itself, and one with it in a package.

Problem 11.2: Overloaded "not" Operator

The NOT operator allows the inversion of binary values. For example, if x = "1000" is a STD_LOGIC_VECTOR value, then NOT x could be used, producing "0111".

However, if x had been declared as an INTEGER, such operation would not be allowed. Write a "not" function capable of inverting integers. (Suggestion: See section 4.4 and example 11.6.)

Problem 11.3:   Logic Shift of std_logic_vector

The pre-defined shift operators (specified in VHDL93, section 4.1) work only with type BIT_VECTOR. Write a function capable of logically shifting a STD_LOGC_ VECTOR signal to the left by a specified amount. Two arguments must be passed to the function: the value to be shifted (STD_LOGIC_VECTOR), plus a NATURAL value specifying the amount of shift. Place your function in a package. Then write an application with a call to your function in order to test it (suggestion: review example 11.7).

Problem 11.4:   Logic Shift of an Integer

This problem is an extension of problem 11.3. Write a function capable of shifting an INTEGER value to the left by an specified amount. Place your function in a package. Then write an application with a call to your function in order to test it.

Problem 11.5:   Signed Multiplier

Write a function similar to that of example 11.8. However, it should now operate with SIGNED input and output values.

Problem 11.6:   Two-digit Counter with SSD Output

In example 6.7, a progressive 2-digit decimal counter ($0 \rightarrow 99 \rightarrow 0$), with external asynchronous reset plus binary-coded decimal (BCD) to seven-segment display (SSD) conversion, was designed. In it, a routine to convert a signal from BCD to SSD format was used twice. This kind of repetition can be avoided with a FUNCTION. Write a function (call it bcd_to_ssd) capable of making such a conversion and place it in a PACKAGE. Then redo the design of example 6.7, using a call to your function whenever such conversion is needed. Then synthesize and test your solution.

Problem 11.7:   Statistical Procedure

Write a PROCEDURE that receives eight signed values and returns their average, the largest value, and the lowest value. Call the return values ave, max, and min. Place your procedure in a package. Then write an application with a call to it in order to test its functionality.

Problem 11.8:   Overloaded " + " Operator

In example 11.6, a function that overloads the "+" (addition) operator was presented. Its purpose was to allow the direct addition of STD_LOGIC_VECTOR values. In that example, the return parameter had the same number of bits as the two input parameters. Write a similar function, but with the return vector having one extra bit corresponding to the carry out bit such that overflow can then be easily detected.

12 Additional System Designs

In this chapter, additional designs are presented, with the purpose of further illustrating the usage of the VHDL units that are intended for system-level design: PACKAGES, COMPONENTS, FUNCTIONS, and PROCEDURES.

12.1   Serial-Parallel Multiplier

Figure 12.1 shows the RTL diagram of a serial-parallel multiplier. One of the input vectors (a) is applied serially to the circuit (one bit at a time, starting from the LSB), while the other (b) is applied in parallel (all bits simultaneously). Say that a has M bits, while b has N. Then, after all M bits of a have been presented to the system, a string of M '0's must follow, in order to complete the (M + N)-bit output product.

As can be seen in figure 12.1, the system is pipelined, and is constructed using AND gates, full-adder units, plus registers (flip-flops). Each unit of the pipeline (except the leftmost one) requires one adder and two registers, plus an AND gate to compute one of the inputs. Thus for an M × N multiplier, O(N) of such units are required.

The solution presented below is of *structural* type (only COMPONENTS were used). Notice that there is more than one level of instantiation (the unit called *pipe* instantiates other components, while in the final code, pipe is instantiated as well (besides other components).

The design of each component is shown below, along with the PACKAGE containing all COMPONENT declarations, followed by the project proper (main code). Simulation results were also included.

```
1   ------ and_2.vhd (component): ---------
2   LIBRARY ieee;
3   USE ieee.std_logic_1164.all;
4   ----------------------------------------
5   ENTITY and_2 IS
6      PORT ( a, b: IN STD_LOGIC;
7              y: OUT STD_LOGIC);
8   END and_2;
9   ----------------------------------------
10  ARCHITECTURE and_2 OF and_2 IS
11  BEGIN
12     y <= a AND b;
13  END and_2;
14  ----------------------------------------
```

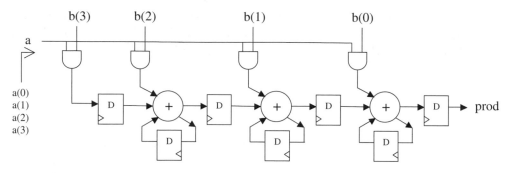

Figure 12.1
Serial-parallel multiplier.

```
1    ------ reg.vhd (component): -----------
2  LIBRARY ieee;
3  USE ieee.std_logic_1164.all;
4  ---------------------------------------
5  ENTITY reg IS
6     PORT ( d, clk, rst: IN STD_LOGIC;
7             q: OUT STD_LOGIC);
8  END reg;
9  ---------------------------------------
10 ARCHITECTURE reg OF reg IS
11 BEGIN
12    PROCESS (clk, rst)
13    BEGIN
14       IF (rst='1') THEN q<='0';
15       ELSIF (clk'EVENT AND clk='1') THEN q<=d;
16       END IF;
17    END PROCESS;
18 END reg;
19 ---------------------------------------

1    ------ fau.vhd (component): -----------
2  LIBRARY ieee;
3  USE ieee.std_logic_1164.all;
4  ---------------------------------------
5  ENTITY fau IS
```

```
6      PORT ( a, b, cin: IN STD_LOGIC;
7              s, cout: OUT STD_LOGIC);
8  END fau;
9  ----------------------------------------
10 ARCHITECTURE fau OF fau IS
11 BEGIN
12    s <= a XOR b XOR cin;
13    cout <= (a AND b) OR (a AND cin) OR (b AND cin);
14 END fau;
15 ----------------------------------------

1  ------ pipe.vhd (component): ---------
2  LIBRARY ieee;
3  USE ieee.std_logic_1164.all;
4  USE work.my_components.all;
5  ----------------------------------------
6  ENTITY pipe IS
7     PORT ( a, b, clk, rst: IN STD_LOGIC;
8             q: OUT STD_LOGIC);
9  END pipe;
10 ----------------------------------------
11 ARCHITECTURE structural OF pipe IS
12    SIGNAL s, cin, cout: STD_LOGIC;
13 BEGIN
14    U1: COMPONENT fau PORT MAP (a, b, cin, s, cout);
15    U2: COMPONENT reg PORT MAP (cout, clk, rst, cin);
16    U3: COMPONENT reg PORT MAP (s, clk, rst, q);
17 END structural;
18 ----------------------------------------

1  ----- my_components.vhd (package):-----
2  LIBRARY ieee;
3  USE ieee.std_logic_1164.all;
4  ----------------------------------------
5  PACKAGE my_components IS
6  -------------------------
7  COMPONENT and_2 IS
8     PORT (a, b: IN STD_LOGIC; y: OUT STD_LOGIC);
```

```
9  END COMPONENT;
10 --------------------------
11 COMPONENT fau IS
12    PORT (a, b, cin: IN STD_LOGIC; s, cout: OUT STD_LOGIC);
13 END COMPONENT;
14 --------------------------
15 COMPONENT reg IS
16    PORT (d, clk, rst: IN STD_LOGIC; q: OUT STD_LOGIC);
17 END COMPONENT;
18 --------------------------
19 COMPONENT pipe IS
20    PORT (a, b, clk, rst: IN STD_LOGIC; q: OUT STD_LOGIC);
21 END COMPONENT;
22 --------------------------
23 END my_components;
24 ---------------------------------------

1  ----- multiplier.vhd (project): -------
2  LIBRARY ieee;
3  USE ieee.std_logic_1164.all;
4  USE work.my_components.all;
5  ---------------------------------------
6  ENTITY multiplier IS
7     PORT ( a, clk, rst: IN STD_LOGIC;
8            b: IN STD_LOGIC_VECTOR (3 DOWNTO 0);
9            prod: OUT STD_LOGIC);
10 END multiplier;
11 ---------------------------------------
12 ARCHITECTURE structural OF multiplier IS
13    SIGNAL and_out, reg_out: STD_LOGIC_VECTOR (3 DOWNTO 0);
14 BEGIN
15    U1: COMPONENT and_2 PORT MAP (a, b(3), and_out(3));
16    U2: COMPONENT and_2 PORT MAP (a, b(2), and_out(2));
17    U3: COMPONENT and_2 PORT MAP (a, b(1), and_out(1));
18    U4: COMPONENT and_2 PORT MAP (a, b(0), and_out(0));
19    U5: COMPONENT reg PORT MAP (and_out(3), clk, rst,
20       reg_out(3));
```

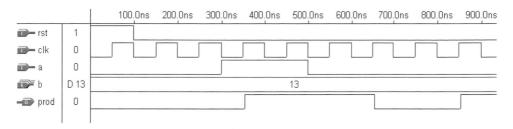

Figure 12.2
Simulation results of serial-parallel multiplier.

```
21     U6: COMPONENT pipe PORT MAP (and_out(2), reg_out(3),
22        clk, rst, reg_out(2));
23     U7: COMPONENT pipe PORT MAP (and_out(1), reg_out(2),
24        clk, rst, reg_out(1));
25     U8: COMPONENT pipe PORT MAP (and_out(0), reg_out(1),
26        clk, rst, reg_out(0));
27     prod <= reg_out(0);
28 END structural;
29 ---------------------------------------
```

Simulation results are shown in figure 12.2. a = "1100" (decimal 12) was applied to the serial input. Notice that this input must start with the LSB (a(0) = '0'), which appears in the time slot 100 ns–200 ns, while the MSB (a(3) = '1') is situated in 400 ns–500 ns. Recall that four zeros must then follow. On the other hand, at the parallel input, b = "1101" (decimal 13) was applied. The expected result, prod = "10011100" (decimal 156), can be observed in the lower plot. Recall that the first bit out is the LSB; that is, prod(0) = '0', which appears in the time slot immediately after the first rising edge of clock; (that is, 150 ns–250 ns), while the last bit (MSB) of prod is situated in 850 ns–950 ns.

12.2   Parallel Multiplier

Figure 12.3 shows the diagram of a 4-bit parallel multiplier. Contrary to the case of figure 12.1, here all input bits are applied to the system simultaneously. Therefore, registers are not required. Notice in figure 12.3 that only AND gates and FAU (full adder units) are necessary to construct a parallel multiplier. The operands are a and b (each of four bits), and the resulting product is prod (eight bits).

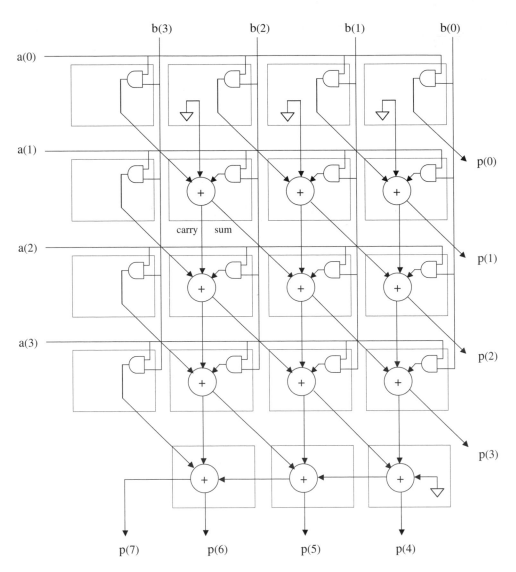

Figure 12.3
Parallel multiplier.

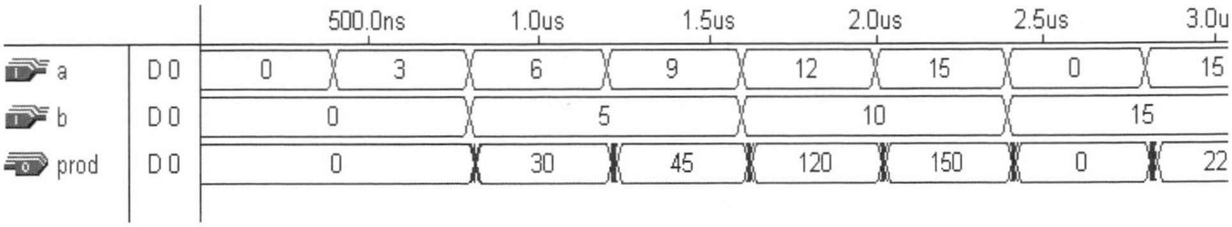

		500.0ns	1.0us	1.5us	2.0us	2.5us	3.0u

Figure 12.4 simulation waveform with signals a, b, prod showing values.

Figure 12.4
Simulation results of parallel multiplier.

The VHDL code shown below was based on COMPONENT instantiation. Notice that two basic components, AND_2 and FAU, were first specified (shown in section 12.1). These components were then instantiated to construct higher-level components, *top_row*, *mid_row*, and *lower_row*. All of these components were then declared in a PACKAGE called *my_components*, and finally used in the project called *multiplier* to implement the circuit of figure 12.3. Simulation results are shown in figure 12.4.

```
1  ------- top_row.vhd (component): -------------
2  LIBRARY ieee;
3  USE ieee.std_logic_1164.all;
4  USE work.my_components.all;
5  ---------------------------------------
6  ENTITY top_row IS
7     PORT ( a: IN STD_LOGIC;
8            b: IN STD_LOGIC_VECTOR (3 DOWNTO 0);
9            sout, cout: OUT STD_LOGIC_VECTOR (2 DOWNTO 0);
10           p: OUT STD_LOGIC);
11 END top_row;
12 ---------------------------------------
13 ARCHITECTURE structural OF top_row IS
14 BEGIN
15    U1: COMPONENT and_2 PORT MAP (a, b(3), sout(2));
16    U2: COMPONENT and_2 PORT MAP (a, b(2), sout(1));
17    U3: COMPONENT and_2 PORT MAP (a, b(1), sout(0));
18    U4: COMPONENT and_2 PORT MAP (a, b(0), p);
19    cout(2)<='0'; cout(1)<='0'; cout(0)<='0';
20 END structural;
21 ---------------------------------------------
```

```
1    ------- mid_row.vhd (component): -------------
2  LIBRARY ieee;
3  USE ieee.std_logic_1164.all;
4  USE work.my_components.all;
5  ----------------------------------------
6  ENTITY mid_row IS
7     PORT ( a: IN STD_LOGIC;
8             b: IN STD_LOGIC_VECTOR (3 DOWNTO 0);
9             sin, cin: IN STD_LOGIC_VECTOR (2 DOWNTO 0);
10            sout, cout: OUT STD_LOGIC_VECTOR (2 DOWNTO 0);
11            p: OUT STD_LOGIC);
12 END mid_row;
13 ----------------------------------------
14 ARCHITECTURE structural OF mid_row IS
15    SIGNAL and_out: STD_LOGIC_VECTOR (2 DOWNTO 0);
16 BEGIN
17    U1: COMPONENT and_2 PORT MAP (a, b(3), sout(2));
18    U2: COMPONENT and_2 PORT MAP (a, b(2), and_out(2));
19    U3: COMPONENT and_2 PORT MAP (a, b(1), and_out(1));
20    U4: COMPONENT and_2 PORT MAP (a, b(0), and_out(0));
21    U5: COMPONENT fau PORT MAP (sin(2), cin(2), and_out(2),
22       sout(1), cout(2));
23    U6: COMPONENT fau PORT MAP (sin(1), cin(1), and_out(1),
24       sout(0), cout(1));
25    U7: COMPONENT fau PORT MAP (sin(0), cin(0), and_out(0),
26       p, cout(0));
27 END structural;
28 --------------------------------------------

1    ------- lower_row.vhd (component): -----------
2  LIBRARY ieee;
3  USE ieee.std_logic_1164.all;
4  USE work.my_components.all;
5  ----------------------------------------
6  ENTITY lower_row IS
7     PORT ( sin, cin: IN STD_LOGIC_VECTOR (2 DOWNTO 0);
8             p: OUT STD_LOGIC_VECTOR (3 DOWNTO 0);
9  END lower_row;
10 ----------------------------------------
```

```
11 ARCHITECTURE structural OF lower_row IS
12    SIGNAL local: STD_LOGIC_VECTOR (2 DOWNTO 0);
13 BEGIN
14    local(0)<='0';
15    U1: COMPONENT fau PORT MAP (sin(0), cin(0), local(0),
16       p(0), local(1));
17    U2: COMPONENT fau PORT MAP (sin(1), cin(1), local(1),
18       p(1), local(2));
19    U3: COMPONENT fau PORT MAP (sin(2), cin(2), local(2),
20       p(2), p(3));
21 END structural;
22 ------------------------------------------------

1  ----- my_components.vhd (package): -----------
2  LIBRARY ieee;
3  USE ieee.std_logic_1164.all;
4  ------------------------------------
5  PACKAGE my_components IS
6     ----------------------
7     COMPONENT and_2 IS
8        PORT ( a, b: IN STD_LOGIC; y: OUT STD_LOGIC);
9     END COMPONENT;
10    ----------------------
11    COMPONENT fau IS    -- full adder unit
12       PORT ( a, b, cin: IN STD_LOGIC; s, cout: OUT STD_LOGIC);
13    END COMPONENT;
14    ----------------------
15    COMPONENT top_row IS
16       PORT ( a: IN STD_LOGIC;
17             b: IN STD_LOGIC_VECTOR (3 DOWNTO 0);
18             sout, cout: OUT STD_LOGIC_VECTOR (2 DOWNTO 0);
19             p: OUT STD_LOGIC);
20    END COMPONENT;
21    ----------------------
22    COMPONENT mid_row IS
23       PORT ( a: IN STD_LOGIC;
24             b: IN STD_LOGIC_VECTOR (3 DOWNTO 0);
25             sin, cin: IN STD_LOGIC_VECTOR (2 DOWNTO 0);
26             sout, cout: OUT STD_LOGIC_VECTOR (2 DOWNTO 0);
```

```
27                 p: OUT STD_LOGIC);
28     END COMPONENT;
29     ----------------------
30     COMPONENT lower_row IS
31        PORT ( sin, cin: IN STD_LOGIC_VECTOR (2 DOWNTO 0);
32                  p: OUT STD_LOGIC_VECTOR (3 DOWNTO 0);
33     END COMPONENT;
34     ----------------------
35 END my_components;
36 ------------------------------------------------

1  ------- multiplier.vhd (project): ------------
2  LIBRARY ieee;
3  USE ieee.std_logic_1164.all;
4  USE work.my_components.all;
5  ----------------------------------------
6  ENTITY multiplier IS
7     PORT ( a, b: IN STD_LOGIC_VECTOR (3 DOWNTO 0);
8             prod: OUT STD_LOGIC_VECTOR (7 DOWNTO 0));
9  END multiplier;
10 ----------------------------------------
11 ARCHITECTURE structural OF multiplier IS
12    TYPE matrix IS ARRAY (0 TO 3) OF
13       STD_LOGIC_VECTOR (2 DOWNTO 0);
14    SIGNAL s, c: matrix;
15 BEGIN
16    U1: COMPONENT top_row PORT MAP (a(0), b, s(0), c(0),
17       prod(0));
18    U2: COMPONENT mid_row PORT MAP (a(1), b, s(0), c(0), s(1),
19       c(1), prod(1));
20    U3: COMPONENT mid_row PORT MAP (a(2), b, s(1), c(1), s(2),
21       c(2), prod(2));
22    U4: COMPONENT mid_row PORT MAP (a(3), b, s(2), c(2), s(3),
23       c(3), prod(3));
24    U5: COMPONENT lower_row PORT MAP (s(3), c(3),
25       prod(7 DOWNTO 4));
26 END structural;
27 ------------------------------------------------
```

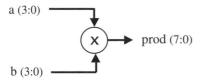

a (3:0)

b (3:0)

x → prod (7:0)

Figure 12.5
Parallel multiplier inferred from the pre-defined "*" operator.

A Simpler Approach

The example above had the purpose of exploring several aspects related to system design using VHDL. However, for the particular case of a parallel multiplier, it can be immediately inferred by means of the pre-defined "*" (multiplication) operator. Therefore, the circuit above can be represented using the compact form of figure 12.5, and the whole code above can be replaced by the following code:

```
1  ---------------------------------------
2  LIBRARY ieee;
3  USE ieee.std_logic_1164.all;
4  USE ieee.std_logic_arith.all;
5  ---------------------------------------
6  ENTITY multiplier3 IS
7     PORT ( a, b: IN SIGNED(3 DOWNTO 0);
8              prod: OUT SIGNED(7 DOWNTO 0));
9  END multiplier3;
10 ---------------------------------------
11 ARCHITECTURE behavior OF multiplier3 IS
12 BEGIN
13    prod <= a * b;
14 END behavior;
15 ---------------------------------------
```

12.3 Multiply-Accumulate Circuits

Multiplication followed by accumulation is a common operation in many digital systems, particularly those highly interconnected, like digital filters, neural networks, data quantizers, etc.

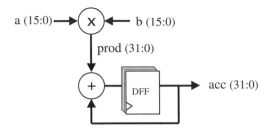

Figure 12.6
MAC circuit.

One typical MAC (multiply-accumulate) architecture is illustrated in figure 12.6. It consists of multiplying two values, then adding the result to the previously accumulated value, which must then be re-stored in the registers for future accumulations. Another feature of a MAC circuit is that it must check for overflow, which might happen when the number of MAC operations is large.

This design can be done using COMPONENTS, because we have already designed each of the units shown in figure 12.6. However, since it is a relatively simple circuit, it can also be designed directly. The latter approach is illustrated below, while the former is treated in problem 12.2. In any case, the MAC circuit, as a whole, can be used as a COMPONENT in applications like digital filters and neural networks (next sections).

Overflow: In the implementation (code) shown below, a FUNCTION was written to detect overflow and truncate the result in case overflow happens. Overflow in a signed adder occurs when two operands with the same signal (leftmost bit) produce a result with a different signal from them. If it occurs, the largest value (positive or negative) should be assigned to the result. For example, if eight bits are used to encode the values, the addition of two positive numbers must fall in the interval from 0 to 127, while the addition of two negative numbers must fall between -128 (that is, $+128$ in unsigned representation) and -1 (255 in unsigned representation). For example, $65 + 65 = 130$, which is indeed -126 (overflow), so the result should be truncated to the largest positive value (127). Likewise, $(-70) + (-70) = -140$, which is, indeed, 116 (overflow), so the result should be truncated to the most negative value (-128). On the other hand, when the operands have different signals, overflow cannot happen.

The add_truncate( ) function was placed in a PACKAGE (chapter 10) called *my_functions*. The function receives two signals, adds them, then checks for overflow and truncates the result if necessary, returning the processed result to the main code. Notice that the function is generic, for the number of bits of the operands is passed to it by means of a parameter called *size*. Notice also in the main code that the parameters passed to the function were declared as signals (line 14), because variables are not allowed (chapter 11).

```
1  ------- PACKAGE my_functions: ---------------------------
2  LIBRARY ieee;
3  USE ieee.std_logic_1164.all;
4  USE ieee.std_logic_arith.all;
5  --------------------------------------------------------
6  PACKAGE my_functions IS
7     FUNCTION add_truncate (SIGNAL a, b: SIGNED; size: INTEGER)
8        RETURN SIGNED;
9  END my_functions;
10 --------------------------------------------------------
11 PACKAGE BODY my_functions IS
12    FUNCTION add_truncate (SIGNAL a, b: SIGNED; size: INTEGER)
13           RETURN SIGNED IS
14       VARIABLE result: SIGNED (7 DOWNTO 0);
15    BEGIN
16       result := a + b;
17       IF (a(a'left)=b(b'left)) AND
18              (result(result'LEFT)/=a(a'left)) THEN
19          result := (result'LEFT => a(a'LEFT),
20                     OTHERS => NOT a(a'left));
21       END IF;
22       RETURN result;
23    END add_truncate;
24 END my_functions;
25 --------------------------------------------------------

1  ------- Main code: ----------------------
2  LIBRARY ieee;
3  USE ieee.std_logic_1164.all;
4  USE ieee.std_logic_arith.all;
5  USE work.my_functions.all;
6  -------------------------------------------
```

```
7   ENTITY mac IS
8      PORT ( a, b: IN SIGNED(3 DOWNTO 0);
9               clk, rst: IN STD_LOGIC;
10              acc: OUT SIGNED(7 DOWNTO 0));
11  END mac;
12  -----------------------------------------
13  ARCHITECTURE rtl OF mac IS
14     SIGNAL prod, reg: SIGNED(7 DOWNTO 0);
15  BEGIN
16     PROCESS (rst, clk)
17        VARIABLE sum: SIGNED(7 DOWNTO 0);
18     BEGIN
19        prod <= a * b;
20        IF (rst='1') THEN
21           reg <= (OTHERS=>'0');
22        ELSIF (clk'EVENT AND clk='1') THEN
23           sum := add_truncate (prod, reg, 8);
24           reg <= sum;
25        END IF;
26     acc <= reg;
27     END PROCESS;
28  END rtl;
29  -----------------------------------------
```

Simulation results are presented in figure 12.7. Notice that the following sequence of signals was presented to the MAC circuit: a = (0, 2, 4, 6, −8, −6, −4, −2), b = (0, 3, 6, −7, −8, −8, −8). Therefore, the expected output sequence is acc = (0, 6, 30, −12, 52, 100, 148) (recall that −12 is represented in the graph as 256 − 12 = 244).

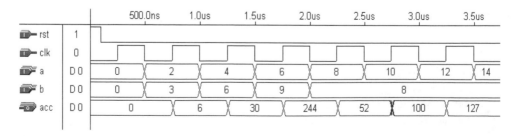

Figure 12.7
Simulation results of MAC circuit.

All the values are OK, except the last one, for it is above the maximum positive value allowed for 8-bit signed numbers (127). Therefore, this result was kept at 127.

12.4   Digital Filters

Digital signal processing (DSP) finds innumerable applications in the fields of audio, video, and communications, among others. Such applications are generally based on LTI (linear time invariant) systems, which can be implemented with digital circuitry.

Any LTI system be represented by the following equation:

$$\sum_{k=0}^{N} a_k y[n-k] = \sum_{k=0}^{M} b_k x[n-k]$$

where a_k and b_k are the filter coefficients, and $x[n-k]$, $y[n-k]$ are the current (for $k=0$) and earlier (for $k>0$) input and output values, respectively. To implement this expression, registers are necessary to store $x[n-k]$ and/or $y[n-k]$ (for $k>0$), besides multipliers and adders, which are well-known building blocks in the digital domain.

The impulse response of a digital filter can be divided into two categories: IIR (infinite impulse response) and FIR (finite impulse response). The former corresponds to the general case described by the equation above, while the latter occurs when $N=0$. Only FIR filters can exhibit linear phase, so they are indispensable when linear phase is required, like in many telecom applications. With $N=0$, the equation above becomes

$$y[n] = \sum_{k=0}^{M} c_k x[n-k]$$

where $c_k = b_k/a_0$ are the coefficients of the FIR filter. This equation can be implemented by the system of figure 12.8, where D (delay) represents a register (flip-flops), a triangle is a multiplier, and a circle means an adder.

An equivalent RTL representation is shown in figure 12.9. As shown, the values of x are stored in a shift register, whose outputs are connected to multipliers and then to adders. The coefficients must also be stored on chip. However, if the coefficients are always the same (that is, if it is a *dedicated* filter), their values can be implemented by means of logic gates rather than registers (we just need to store CONSTANTS). On the other hand, if it is a *general purpose* filter, then registers are required for the coefficients. In the architecture of figure 12.9, the output vector (y) was also stored, in order to provide a clean, synchronous output.

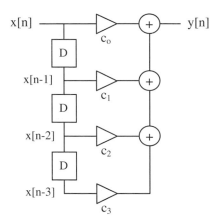

Figure 12.8
FIR filter diagram (with 4 coefficients).

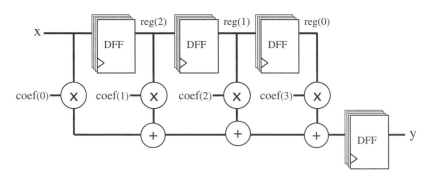

Figure 12.9
RTL representation of a FIR filter.

The circuit of figure 12.9 can be constructed in several ways. However, if it is intended for future reuse or sharing, than it should be as *generic* as possible. In the code presented below, two GENERIC parameters are specified (line 7): n defines the number of filter coefficients, while m specifies the number of bits used to represent the input and coefficients. For the output, 2 m bits were used. Thus, for example, 16 bits could be used for x, coef, and reg, while 32 bits could be used for all other signals (from the outputs of the multipliers all the way to y).

Notice that the lower section of the filter contains a MAC (multiply-accumulate) pipeline. This circuit is closely related to the MAC circuit discussed in section 12.3. Here too, overflow can happen, so an add/truncate procedure must be included in the design.

In the solution below, the coefficients were considered as CONSTANTS (line 19), thus inferring no flip-flops. The values chosen were coef(0) = 4, coef(1) = 3, coef(2) = 2, and coef(3) = 1. Small values were chosen for n and m (4 for both) in order to make the simulation results easy to visualize. With n = m = 4, the synthesized circuit required 20 flip-flops (four for each stage of the shift register, plus eight for the output). As described in chapter 7, flip-flops are inferred when a signal assignment is made on the transition of another signal, which occurs in lines 33–45 of the code below (notice that indeed VARIABLE assignments are made in lines 33–38, but since their values are then passed to a SIGNAL (y), registers are inferred).

```
1  -------------------------------------------------------
2  LIBRARY ieee;
3  USE ieee.std_logic_1164.all;
4  USE ieee.std_logic_arith.all;  -- package needed for SIGNED
5  -------------------------------------------------------
6  ENTITY fir2 IS
7     GENERIC (n: INTEGER := 4; m: INTEGER := 4);
8     -- n = # of coef., m = # of bits of input and coef.
9     -- Besides n and m, CONSTANT (line 19) also need adjust
10    PORT ( x: IN SIGNED(m-1 DOWNTO 0);
11           clk, rst: IN STD_LOGIC;
12           y: OUT SIGNED(2*m-1 DOWNTO 0));
13 END fir2;
14 -------------------------------------------------------
15 ARCHITECTURE rtl OF fir2 IS
16    TYPE registers IS ARRAY (n-2 DOWNTO 0) OF
17                             SIGNED(m-1 DOWNTO 0);
18    TYPE coefficients IS ARRAY (n-1 DOWNTO 0) OF
19                             SIGNED(m-1 DOWNTO 0);
20    SIGNAL reg: registers;
21    CONSTANT coef: coefficients := ("0001", "0010", "0011",
22                                    "0100");
23 BEGIN
24    PROCESS (clk, rst)
25       VARIABLE acc, prod:
26          SIGNED(2*m-1 DOWNTO 0) := (OTHERS=>'0');
27       VARIABLE sign: STD_LOGIC;
28    BEGIN
29       ----- reset: -------------------------
```

```
30          IF (rst='1') THEN
31             FOR i IN n-2 DOWNTO 0 LOOP
32                FOR j IN m-1 DOWNTO 0 LOOP
33                   reg(i)(j) <= '0';
34                END LOOP;
35             END LOOP;
36          ----- register inference + MAC: -------
37          ELSIF (clk'EVENT AND clk='1') THEN
38             acc := coef(0)*x;
39             FOR i IN 1 TO n-1 LOOP
40                sign := acc(2*m-1);
41                prod := coef(i)*reg(n-1-i);
42                acc := acc + prod;
43                ---- overflow check: ------------
44                IF (sign=prod(prod'left)) AND
45                      (acc(acc'left) /= sign)
46                   THEN
47                   acc := (acc'LEFT => sign, OTHERS => NOT sign);
48                END IF;
49             END LOOP;
50             reg <= x & reg(n-2 DOWNTO 1);
51          END IF;
52          y <= acc;
53       END PROCESS;
54 END rtl;
55 -----------------------------------------------------------
```

Simulation results are shown in figure 12.10. Recall that the coefficients are $coef(0) = 4$, $coef(1) = 3$, $coef(2) = 2$, and $coef(3) = 1$, and that the numbers are

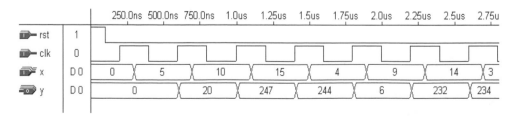

Figure 12.10
Simulation results of FIR filter of figure 12.9.

SIGNED (therefore, with 4-bit values, the range is from -8 to $+7$). The sequence applied to the input was $x[0] = 0$, $x[1] = 5$, $x[2] = -6$ ($16 - 6 = 10$ in the graph), $x[3] = -1$ ($16 - 1 = 15$ in the graph), $x[4] = 4$, $x[5] = -7$ ($16 - 7 = 9$ in the graph), and $x[6] = -2$ ($16 - 2 = 14$ in the graph). Therefore, with all flip-flops previously reset, at the first positive edge of clk the expected output is $y[0] = coef(0)*x[0] = 0$, which coincides with the first result for y in figure 12.10. At the next upward transition of clk, the expected value is $y[1] = coef(0)*x[1] + coef(1)*x[0] = 20$. And one clock cycle later, $y[1] = coef(0)*x[2] + coef(1)*x[1] + coef(2)*x[0] = -9$ ($256 - 9 = 247$ in the graph), and so on.

General Purpose FIR Filter

The design presented above contained fixed coefficients, and is therefore adequate for an ASIC with a dedicated filter. For a general purpose implementation (that is, with *programmable* coefficients), the architecture of figure 12.11 can be used instead. As can be seen, this structure is modular and allows several chips to be cascaded, which might be helpful in some applications, because FIR filters tend to have many taps (coefficients).

In this structure, there are two shift registers, one for storing the inputs (x) and the other for the coefficients (coef). The structure is divided into n equal modules, called TAP1,..., TAPn. Each module (TAP) contains a slice of the shift registers, plus a multiplier and an adder. It also contains an output register, but this is optional (could be used at the last TAP only). This would, however, increase the ripple propagation

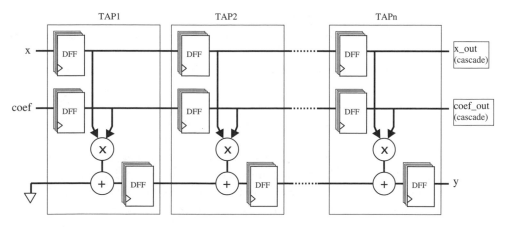

Figure 12.11
General purpose FIR filter.

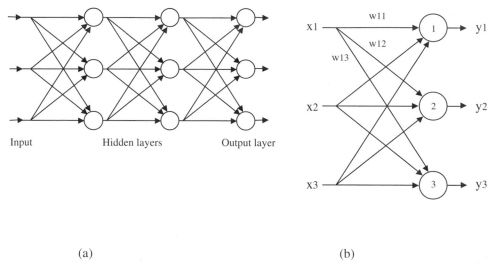

(a)   (b)

Figure 12.12
Feedforward neural network.

between the adders. Of course, all coefficients must be loaded before the computation starts. This FIR architecture will be object of problem 12.4.

12.5   Neural Networks

Neural Networks (NN) are highly parallel, highly interconnected systems. Such characteristics make their implementation very challenging, and also very costly, due to the large amount of hardware required.

A feedforward NN is shown in figure 12.12(a). In this example, the circuit has three layers, with three 3-input neurons in each layer. Internal details of each layer are depicted in figure 12.12(b). xi represents the ith input, wij is the weight between input i and neuron j, and yj is the jth output. Therefore, y1 = f(x1.w11 + x2.w21 + x3.w31), y2 = f(x1.w12 + x2.w22 + x3.w32), and y3 = f(x1.w13 + x2.w23 + x3.w33), where f( ) is the activation function (linear threshold, sigmoid, etc.).

A "ring" architecture for the NN of figure 12.12 is presented in figure 12.13, which implements one layer of the NN. Each box represents one neuron. As shown, there are several *circular* shift registers, one for each neuron (vertical shifters) plus one for the whole set (horizontal shifter). The vertical shifters hold the weights, while the horizontal one holds the inputs (shift registers with 'data_load' capability). Notice

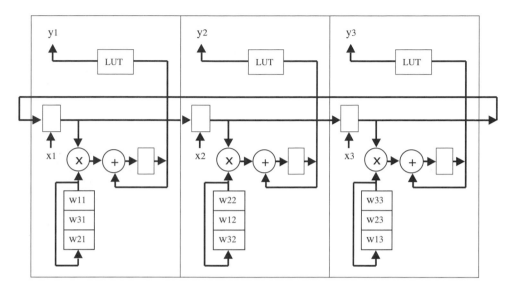

Figure 12.13
Ring architecture for NN implementation.

that the relative position of the weights in their respective registers must match that of the input values. At the output of a vertical shifter there is a MAC circuit (section 12.3), which accumulates the product between the weights and the inputs. All shifters use the same clock signal. Therefore, after one complete circulation, the following values will be available at the output of the MAC circuits: $x1.w11 + x2.w21 + x3.w31$, $x1.w12 + x2.w22 + x3.w32$, and $x1.w13 + x2.w23 + x3.w33$. These values are then applied to a LUT (lookup table), which implements the activation function (sigmoid, for example), thus producing the actual outputs, yi, of the NN.

In this kind of circuit, truncation must be considered. Say that the inputs and weights are 16 bits long. Then at the output of the MAC cells 32-bit numbers would be the natural choice. However, since the actual outputs (after the LUT) might be connected to another layer of neurons, truncation to 16 bits is required. This can be done in the LUT or in the MAC circuit.

Another approach is presented in figure 12.14, which is appropriate for general-purpose NNs (that is, with programmable weights). It employs only one input to load all weights (thus saving on chip pins). In figure 12.14, the weights are shifted in sequentially until each register is loaded with its respective weight. The weights are then multiplied by the inputs and accumulated to produce the desired outputs.

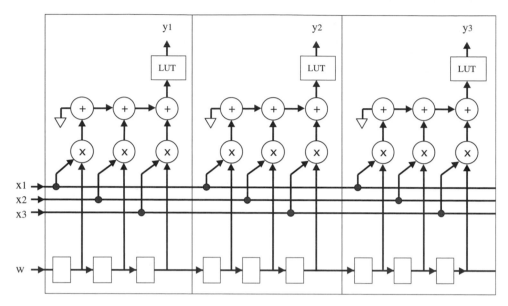

Figure 12.14
NN implementation with only one input for the weights.

Two VHDL codes are presented below, both implementing the architecture of figure 12.14. However, in both solutions the LUT was not included (this will be treated in problem 12.5). The main difference between these two solutions is that the first code is not as *generic*, and is therefore adequate for specific, small designs. The second solution, being generic, is reusable and easily adaptable to any NN size.

Solution 1:   For Small Neural Networks

The solution below has the advantage of being simple, easily understandable, and self-contained in the main code. Its only limitation is that the inputs (x) and outputs (y) are specified one by one rather than using some kind of two-dimensional array, thus making it inappropriate for large NNs. Everything else is generic.

```
1  ------------------------------------------------------------
2  LIBRARY ieee;
3  USE ieee.std_logic_1164.all;
4  USE ieee.std_logic_arith.all;   -- package needed for SIGNED
5  ------------------------------------------------------------
```

```
6  ENTITY nn IS
7  GENERIC ( n: INTEGER := 3;   -- # of neurons
8             m: INTEGER := 3; -- # of inputs or weights per neuron
9             b: INTEGER := 4); -- # of bits per input or weight
10 PORT ( x1: IN SIGNED(b-1 DOWNTO 0);
11         x2: IN SIGNED(b-1 DOWNTO 0);
12         x3: IN SIGNED(b-1 DOWNTO 0);
13         w: IN SIGNED(b-1 DOWNTO 0);
14         clk: IN STD_LOGIC;
15         test: OUT SIGNED(b-1 DOWNTO 0);   -- register test output
16         y1: OUT SIGNED(2*b-1 DOWNTO 0);
17         y2: OUT SIGNED(2*b-1 DOWNTO 0);
18         y3: OUT SIGNED(2*b-1 DOWNTO 0));
19 END nn;
20 -------------------------------------------------------------
21 ARCHITECTURE neural OF nn IS
22    TYPE weights IS ARRAY (1 TO n*m) OF SIGNED(b-1 DOWNTO 0);
23    TYPE inputs IS ARRAY (1 TO m) OF SIGNED(b-1 DOWNTO 0);
24    TYPE outputs IS ARRAY (1 TO m) OF SIGNED(2*b-1 DOWNTO 0);
25 BEGIN
26    PROCESS (clk, w, x1, x2, x3)
27       VARIABLE weight: weights;
28       VARIABLE input: inputs;
29       VARIABLE output: outputs;
30       VARIABLE prod, acc: SIGNED(2*b-1 DOWNTO 0);
31       VARIABLE sign: STD_LOGIC;
32    BEGIN
33       ----- shift register inference: -------------
34       IF (clk'EVENT AND clk='1') THEN
35          weight := w & weight(1 TO n*m-1);
36       END IF;
37       --------- initialization: --------------------
38       input(1) := x1;
39       input(2) := x2;
40       input(3) := x3;
41       ------ multiply-accumulate: -----------------
42       L1: FOR i IN 1 TO n LOOP
43          acc := (OTHERS => '0');
```

```
44              L2: FOR j IN 1 TO m LOOP
45                  prod := input(j)*weigth(m*(i-1)+j);
46                  sign := acc(acc'LEFT);
47                  acc := acc + prod;
48                  ---- overflow check: -----------------
49                  IF (sign=prod(prod'left)) AND
50                       (acc(acc'left) /= sign) THEN
51                     acc := (acc'LEFT => sign, OTHERS => NOT sign);
52                  END IF;
53              END LOOP L2;
54              output(i) := acc;
55          END LOOP L1;
56          --------- outputs: -------------------------
57          test <= weight(n*m);
58          y1 <= output(1);
59          y2 <= output(2);
60          y3 <= output(3);
61      END PROCESS;
62 END neural;
63 ----------------------------------------------------------------
```

Simulation results are shown in figure 12.15. Notice that a small number of bits and a small quantity of neurons were used in order to ease the visualization of the simulation results. As can be seen in lines 7–9 of the code above, the NN has three neurons with three 4-bit inputs each. Since type SIGNED was employed, the range

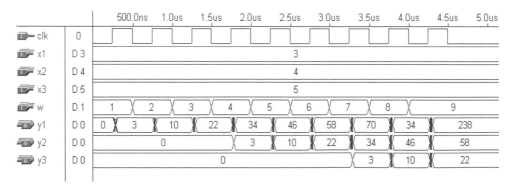

Figure 12.15
Simulation results of NN implemented in solution 1.

of the input values and weights runs from -8 to 7, and the range of the outputs (8 bits) runs from -128 to 127. The inputs were kept fixed at $x1 = 3$, $x2 = 4$, and $x3 = 5$. Since there are nine weights, nine clock cycles are needed to shift them in, as shown in figure 12.5. The values chosen for the weights were $w9 = 1$, $w8 = 2, \ldots,$ $w1 = 9$ (notice that the first weight in is indeed $w9$, for it is shifted nine positions over). Recall, however, that 9 is indeed -7, and 8 is -8, because our data type is SIGNED. Therefore, after the weights have been all loaded, the system immediately furnishes its first set of outputs; that is: $y1 = x1.w1 + x2.w2 + x3.w3 = (3)(-7) + (4)(-8) + (5)(7) = -18$ (represented as $256 - 18 = 238$); $y2 = x1.w4 + x2.w5 + x3.w6 = (3)(6) + (4)(5) + (5)(4) = 58$; and $y3 = x1.w7 + x2.w8 + x3.w9 = (3)(3) + (4)(2) + (5)(1) = 22$. These values (238, 58, and 22) can be seen at the right end of figure 12.15.

Solution 2:   For Large Neural Networks

The code below is generic. Moreover, the inputs and outputs were declared as two-dimensional arrays (section 3.5), thus easily allowing the construction of NNs of any size.

To specify the arrays needed in the design, a PACKAGE named *my_data_types* was employed. As can be seen, it contains two user-defined data types, *vector_array_in* and *vector_array_out*. The PACKAGE was then made visible to the design by means of a USE clause (line 5 of the main code). In this way, the new data types are truly global, and so can be used even in the ENTITY of the main code (that is, in the specification of PORT). These data types were used to specify the inputs and outputs of the systems (lines 11 and 15, respectively). Therefore, all parameters are now generic and easily modifiable, regardless of the size of the NN to be constructed.

Notice in the code below that this solution was divided into two very short parts: *sequential logic* (shift register implementation) in lines 26–28, followed by *combinational logic* (MAC) implementation. A *test* output (for checking the last register) was also included, which is obviously optional. As in all our previous MAC circuit implementations, a routine to check for overflow was also included (lines 39–41).

```
1    -------- Package my_data_types: ---------------------------
2    LIBRARY ieee;
3    USE ieee.std_logic_1164.all;
4    USE ieee.std_logic_arith.all;   -- package needed for SIGNED
5    ----------------------------
6    PACKAGE my_data_types IS
7       CONSTANT b: INTEGER := 3;   -- # of bits per input or weight
```

```
8     TYPE vector_array_in IS ARRAY (NATURAL RANGE <>) OF
9                               SIGNED(b-1 DOWNTO 0);
10    TYPE vector_array_out IS ARRAY (NATURAL RANGE <>) OF
11                              SIGNED(2*b-1 DOWNTO 0);
12 END my_data_types;
13 ------------------------------------------------------------

1  --------- Project nn (main code): -------------------------
2  LIBRARY ieee;
3  USE ieee.std_logic_1164.all;
4  USE ieee.std_logic_arith.all;  -- package needed for SIGNED
5  USE work.my_data_types.all;    -- package of user-defined types
6  ------------------------------------------------------------
7  ENTITY nn3 IS
8     GENERIC ( n: INTEGER := 3;   -- # of neurons
9               m: INTEGER := 3;   -- # of inputs or weights per
10                                 -- neuron
11              b: INTEGER := 3);  -- # of bits per input or
12                                 -- weight
13    PORT ( x: IN VECTOR_ARRAY_IN (1 TO m);
14           w: IN SIGNED(b-1 DOWNTO 0);
15           clk: IN STD_LOGIC;
16           test: OUT SIGNED(b-1 DOWNTO 0);  -- register test
17                                            -- output
18           y: OUT VECTOR_ARRAY_OUT(1 TO n));
19 END nn3;
20 ------------------------------------------------------------
21 ARCHITECTURE neural OF nn3 IS
22 BEGIN
23    PROCESS (clk, w, x)
24       VARIABLE weight: VECTOR_ARRAY_IN (1 TO m*n);
25       VARIABLE prod, acc: SIGNED(2*b-1 DOWNTO 0);
26       VARIABLE sign: STD_LOGIC;
27    BEGIN
28       ----- shift register inference: --------------
29       IF (clk'EVENT AND clk='1') THEN
30          weight := w & weight(1 TO n*m-1);
31       END IF;
```

```
32         test <= weight(n*m);
33         ---- initialization: ------------------------
34         acc := (OTHERS => '0');
35         ------ multiply-accumulate: ------------------
36         L1: FOR i IN 1 TO n LOOP
37            L2: FOR j IN 1 TO m LOOP
38               prod := x(j)*weight(m*(i-1)+j);
39               sign := acc(acc'LEFT);
40               acc := acc + prod;
41               ---- overflow check: ------------------
42               IF (sign=prod(prod'LEFT)) AND
43                     (acc(acc'LEFT)/=sign) THEN
44                  acc := (acc'LEFT => sign, OTHERS => NOT sign);
45               END IF;
46            END LOOP L2;
47            ------ output: ------------------------
48            y(i) <= acc;
49            acc := (OTHERS => '0');
50         END LOOP L1;
51      END PROCESS;
52 END neural;
53 ----------------------------------------------------------------
```

Other aspects related to the design of NNs will be treated in problem 12.5.

12.6   Problems

This section contains a series of problems regarding the use of system-level VHDL units (PACKAGES, COMPONENTS, FUNCTIONS, and PROCEDURES).

Problem 12.1:   Parallel Multiplier

We have seen, in section 12.2, the implementation of a parallel multiplier from scratch. It was also mentioned that the pre-defined "*" (multiplication) operator implements a parallel multiplier too. Though there are several architectures for such a circuit (one was shown in figure 12.3), it is reasonable to assume that the amount of hardware necessary to implement either solution presented in section 12.2 (from scratch or using "*") should not differ substantially. You are asked to synthesize

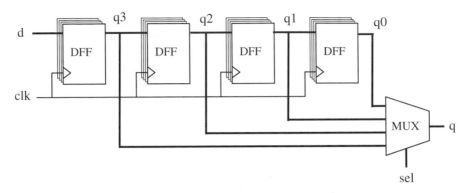

Figure P12.2.

both solutions and compare the resulting report files. Choose several PLD/FPGA target chips. What is the number of product terms and logic cells required in each case? Are their quantities of the same order?

Problem 12.2:   Shifter

Consider the 4-stage shift register of figure P12.2, whose actual output (q) is selected by means of a multiplexer. Say that the data bus is eight-bit wide (thus each register is composed of eight D-type flip-flops).

(a) Create two COMPONENTS, *reg* and *mux*, and then make use of them to construct the complete circuit of figure P12.2.

(b) Assume now that we want to implement only the shift register, without the multiplexer, but that all registered values (q0, q1, q2, and q3) must be available at the output. Write a VHDL code for such a circuit.

(c) Let us consider the same situation of (b) above. However, we now want the design to be *generic* (that is, to have *n* stages, and *b* bits per stage, with such parameters specified by means of a GENERIC statement). In this case, an user-defined array will be necessary to specify the outputs (call the outputs *qout*). Write such a code. (Suggestion: review section 3.5 and/or examine the second design of section 12.5).

(d) Finally, in continuation to the design of (c) above, assume that we want to add 'data load' capability to the shift register. Add an extra input (call it *x*) to each register and an extra pin to (call it *load*), such that when load is asserted all registers are overwritten with the values presented at the inputs. For x, the same user-defined TYPE created for qout can (and should) be used.

Problem 12.3: MAC Circuit

In section 12.3, we studied the implementation of a MAC (multiply-accumulate) circuit (figure 12.6). In the implementation shown there, a FUNCTION was employed, but COMPONENTS were not. Write another solution, this time using COMPONENTS (multiplier, adder, and register). Create the components, then instantiate them in the main code. Compile and simulate your project, comparing your results with those obtained in figure 12.7

Problem 12.4: General Purpose FIR Filter

In section 12.4, we discussed the implementation of FIR filters. One complete design was presented, in which the coefficients of the filter were fixed (figure 12.9). For a general purpose filter (programmable coefficients), a modular architecture was suggested in figure 12.11. You are asked to write a VHDL code for that filter. As a suggestion, review first sections 12.3 and 12.4. Do not forget to include overflow check in your design. Consider that the number of bits of all signals from the input (x and coef) up to the multiplier inputs is m, and 2m from there on (that is, from the multiplier outputs up to y). Consider also that the number of taps (stages) is n. Write a code as generic as possible. Then synthesize and simulate your circuit.

Problem 12.5: Neural Network

In section 12.5, we discussed the implementation of a highly interconnected system: a neural network. Two architectures were presented, and two VHDL codes were written regarding the second architecture. However, the LUT was not included in those solutions. In this problem, the following is asked:

(a) Write a VHDL code that implements a LUT (you can choose the function to be implemented, because what we want to practice here is how to implement a LUT). Recall that a lookup table is simply a ROM (section 9.10).

(b) Write a VHDL code that implements the neural architecture depicted in figure 12.13. Then synthesize and simulate your solution to verify whether it works as expected.

(c) There certainly are other ways of implementing a NN besides the two approaches presented in section 12.5. Can you suggest another one? Can you suggest improvements on the architectures and solutions presented there?

Appendix A: Programmable Logic Devices

A1. Introduction

Programmable Logic Devices (PLDs) were introduced in the mid 1970s. The idea was to construct combinational logic circuits that were *programmable*. However, contrary to microprocessors, which can *run* a program but posses a *fixed* hardware, the programmability of PLDs was intended at the *hardware* level. In other words, a PLD is a *general purpose* chip whose *hardware* can be reconfigured to meat particular specifications.

The first PLDs were called PAL (Programmable Array Logic) or PLA (Programmable Logic Array), depending on the programming scheme (discussed later). They used only logic gates (no flip-flops), thus allowing only the implementation of *combinational* circuits. To circumvent this problem, *registered* PLDs were launched soon after, which included one flip-flop at each output of the circuit. With them, simple *sequential* functions could then be implemented as well.

In the beginning of the 1980s, additional logic circuitry was added to each PLD output. The new output cell, called Macrocell, contained (besides the flip-flop) logic gates and multiplexers. Moreover, the cell itself was programmable, allowing several modes of operation. Additionally, it provided a 'return' (feedback) signal from the output of the circuit to the programmable array, which gave the PLD greater flexibility. This new PLD structure was called *generic* PAL (GAL). A similar architecture was known as PALCE (PAL CMOS Electrically erasable/programmable) device.

All these chips (PAL, PLA, registered PLD, and GAL/PALCE) are now collectively referred to as SPLDs (Simple PLDs). The GAL/PALCE device is the only still manufactured in a standalone package.

Later, several GAL devices were fabricated on the same chip, using a more sophisticated routing scheme, more advanced silicon technology, and several additional features (like JTAG support and interface to several logic standards). This approach became known as CPLD (Complex PLD). CPLDs are currently very popular due to their high density, high performance, and low cost (CPLDs under a dollar can be found).

Finally, in the mid 1980s, FPGAs (Field Programmable Gate Arrays) were introduced. FPGAs differ from CPLDs in architecture, technology, built-in features, and cost. They are aimed mainly at the implementation of large size, high-performance circuits.

A summary of the evolution of PLDs is presented in the table below.

PLDs	Simple PLD (SPLD)	PAL PLA Registered PAL/PLA GAL
	Complex PLD (CPLD)	
	FPGA	

A final remark: all PLDs (simple or complex) are non-volatile. They can be OTP (one-time programmable), in which case fuses or antifuses are used, or can be reprogrammable, with EEPROM or Flash memory (Flash is the technology of choice in most new devices). FPGAs, on the other hand, are mostly volatile, for they make use of SRAM to store the connections, in which case a configuration ROM is necessary to load the interconnects at power up. There are, however, non-volatile options, like the use of antifuse. Examples of each alternative will be shown later.

A2.   SPLDs (Simple PLDs)

As mentioned above, PAL, PLA, and GAL devices are collectively called Simple PLDs (SPLDs). A description of each of these architectures follows.

PAL Devices

PAL (Programmable Array Logic) chips were introduced by Monolithic Memories in the mid 1970s. Its basic architecture is illustrated symbolically in figure A1, where the little circles represent programmable connections. As can be seen, the circuit is composed of a *programmable* array of AND gates, followed by a *fixed* array of OR gates.

The implementation of figure A1 was based on the fact that any combinational function can be represented by a Sum-of-Products (SOP); that is, if $a_1, a_2, \ldots, a_N$ are the logic inputs, then any combinational output x can be computed as

$$x = m_1 + m_2 + \cdots + m_M,$$

where $m_i = f_i (a_1, a_2, \ldots, a_N)$ are the minterms of the function x. For example

$$x = a_1 \bar{a}_2 + a_2 a_3 \bar{a}_4 + \bar{a}_1 \bar{a}_2 a_3 a_4 \bar{a}_5.$$

inputs

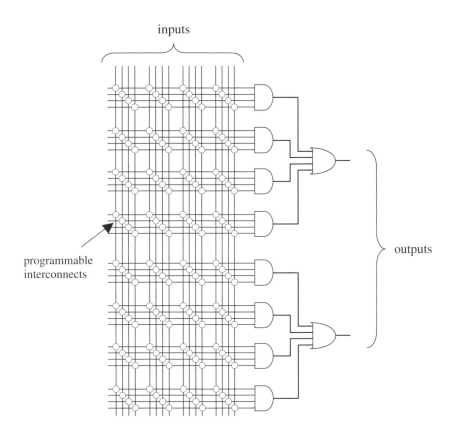

programmable
interconnects

outputs

Figure A1
Illustration of PAL architecture.

Hence, the products (minterms) can be obtained by means of AND gates, whose outputs are then connected to an OR gate to compute their sum, thus implementing the SOP equation described above.

The main limitation of this approach was the fact that it allowed only the implementation of combinational functions. To circumvent this problem, *registered* PALs were launched toward the end of the 1970s. These included a flip-flop at each output (after the OR gates in figure A1), thus allowing the implementation of *sequential* functions as well (though only very simple ones).

An example of a then popular PAL chip is the PAL16L8 device, which contained 16 inputs and 8 outputs (though only 18 I/O pins were indeed available, because it was a 20-pin DIP package; there were ten IN pins, two OUT pins, and six IN/OUT

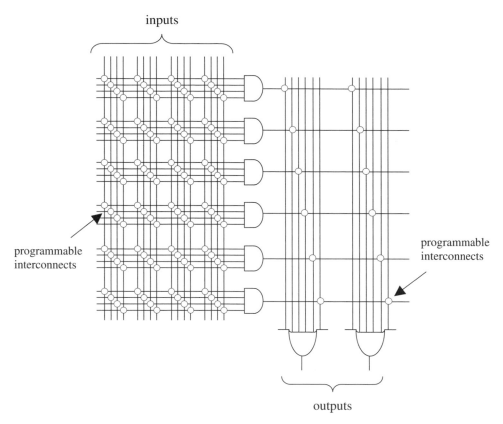

inputs

programmable
interconnects

programmable
interconnects

outputs

Figure A2
Illustration of PLA architecture.

pins (bidirectional), plus VCC and GND). Its *registered* counterpart was the 16R8 chip (where R stands for Registered).

The early technology employed in the fabrication of PAL devices was bipolar, with 5 V supply and current consumption (with open outputs) around 200 mA. The maximum frequency was of the order of 100 MHz, and the programmable cells were of PROM (fuse links) or EPROM (20min UV erase time) type.

PLA Devices

PLA (Programmable Logic Array) chips were also introduced in the mid 1970s (by Signetics). The basic architecture of a PLA is illustrated symbolically in figure A2. Comparing it with figure A1, we observe that the only fundamental difference between them is that while a PAL has programmable AND connections and fixed OR connections, *both* are programmable in a PLA. The obvious advantage was greater flexibility. However, higher time constants at the internal nodes lowered the circuit speed.

An example of a then popular PLA chip is the Signetics PLS161 device. It contained 12 inputs and 8 outputs, being the AND inputs and the OR inputs all programmable. A total of 48 12-input AND gates were available, followed by a total of 8 48-input OR gates. At the outputs, additional programmable XOR gates were also available.

The technology then employed in the fabrication of PLAs was the same as that of PALs. Though PLAs are also obsolete now, they reappeared recently as a building block in the first family of low power CPLDs, the CoolRunner family (from Xilinx—to be described later).

GAL Devices

The GAL (Generic PAL) architecture was introduced by Lattice in the beginning of the 1980s. It contained several important improvements over the first PAL devices: first, a more sophisticated output cell (Macrocell) was constructed, which included, besides the flip-flop, several gates and multiplexers; second, the Macrocell itself was programmable, allowing several modes of operation; third, a 'return' signal from the output of the Macrocell to the programmable array was also included, conferring the circuit more versatility; fourth, EEPROM was employed instead of PROM or EPROM. An electronic signature for identification was also included.

As mentioned earlier, GAL is the only SPLD (Simple PLD) still manufactured in a standalone package. Additionally, it also serves as the basic building block in the construction of most CPLDs (there are exceptions, however, like the CoolRunner CPLD mentioned above, which employs PLAs instead).

Figure A3 shows an example of GAL device, the GAL16V8 (where V stands for Versatile). It is a 16-input, 8-output circuit in a 20-pin package. As can be seen, the actual configuration is eight IN pins (pis 2–9) and eight IN/OUT pins (pins 12–19), plus CLK (pin 1), /OE (–Output Enable, pin 11), VDD (pin 20), and GND (pin 10). At each output there is a Macrocell (after the OR gate), which contains, besides the flip-flop, logic gates and multiplexers. A feedback signal from the Macrocell to the programmable array can also be observed. The programmable interconnections are represented by small circles. Notice that this architecture directly resembles that of a PAL (figure A1), except for the presence of a macrocell at each output and the feedback signal.

Current GAL devices use CMOS technology, 3.3 V supply, EEPROM or Flash technology, and maximum frequency around 250 MHz. Several companies manufacture them (Lattice, Atmel, TI, etc.).

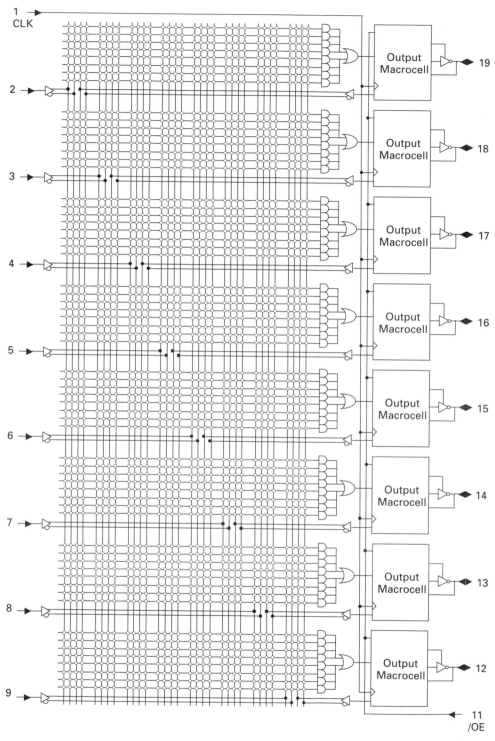

Figure A3
GAL 16V8 chip.

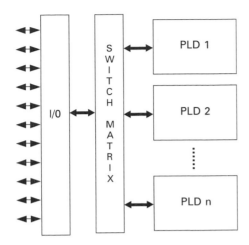

Figure A4
CPLD architecture.

A3. CPLD (Complex PLD)

The basic approach in the construction of a CPLD is illustrated in figure A4. As shown, it consists of several PLDs (in general of GAL type) fabricated on a single chip, with a programmable switch matrix used to connect them together and to the I/O pins. Moreover, CPLDs normally contain a few additional features, like JTAG support and interface to other logic standards (1.8 V, 2.5 V, 5 V, etc.).

Regarding figure A4, as an example we can mention the Xilinx XC9500 CPLD. It consists of n PLDs, each resembling a 36V18 GAL device (therefore similar to the 16V8 architecture of figure A3, but with 36 inputs and 18 outputs, instead of 16 inputs and 8 outputs, thus with 18 Macrocells each), where n = 2, 4, 6, 8, 12, or 16.

Several companies manufacture CPLDs, like Altera, Xilinx, Lattice, Atmel, Cypress, etc. Examples from two companies (Altera and Xilinx) are illustrated in tables A1 and A2. As can be seen, over 500 macrocells and over 10,000 gates can be found in these devices.

A4. FPGA

Field Programmable Gate Array (FPGA) devices were introduced by Xilinx in the mid 1980s. They differ from CPLDs in architecture, storage technology, number of built-in features, and cost, and are aimed at the implementation of high performance, large-size circuits.

Table A1
Altera CPLDs.

Family	Max7000 (B, AE, S)	MAX3000 (A)	MAX II (G)
Macrocells/LUTs	32–512 macrocells	32–512 macrocells	240–2,210 LUTs (192–1,700 equiv. macrocells)
System gates	600–10,000	600–10,000	
I/O pins	32–512	34–208	80–272
Max. internal clock freq.	303 MHz	227 MHz	304 MHz (I/O limited)
Supply voltage	2.5 V (B), 3.3 V (AE), 5 V (S)	3.3 V	1.8 V (G), 2.5 V, 3.3 V
Interconnects	EEPROM	EEPROM	Flash + SRAM
Static current	9 mA–450 mA	9 mA–150 mA	2 mA–50 mA
Technology	0.22 u CMOS EEPROM 4-layer metal (7000 B)	0.3 u, 4-layer metal	0.18 u, 6-layer metal

Table A2
Xilinx CPLDs.

Family	XC9500 (XV, XL, −)	CoolRunner XPLA3	CoolRunner II
Macrocells	36–288	32–512	32–512
System gates	800–6,400	750–12,000	750–12,000
I/O pins	34–192	36–260	33–270
Max. internal clock frequency	222 MHz	213 MHz	385 MHz
Building block	GAL 54V18 (XV, XL) GAL 36V18 (−)	PLA block	PLA block
Supply voltage	2.5 V (XV), 3.3 V (XL), 5 V	3.3 V	1.8 V
Interconnects	Flash	EEPROM	
Technology	0.35 u CMOS	0.35 u CMOS	0.18 u CMOS
Static current	11–500 mA	<0.1 mA	22 uA–1 mA

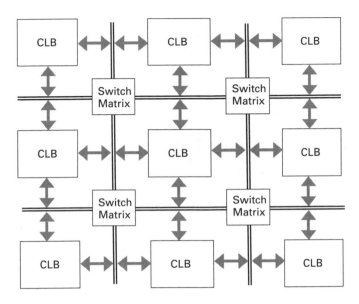

Figure A5
FPGA architecture.

The basic architecture of an FPGA is illustrated in figure A5. It consists of a matrix of CLBs (Configurable Logic Blocks), interconnected by an array of switch matrices.

The internal architecture of a CLB (figure A5) is different from that of a PLD (figure A4). First, instead of implementing SOP expressions with AND gates followed by OR gates (like in SPLDs), its operation is normally based on a LUT (lookup table). Moreover, in an FPGA the number of flip-flops is much more abundant than in a CPLD, thus allowing the construction of more sophisticated sequential circuits. Besides JTAG support and interface to diverse logic levels, other additional features are also included in FPGA chips, like SRAM memory, clock multiplication (PLL or DLL), PCI interface, etc. Some chips also include dedicated blocks, like multipliers, DSPs, and microprocessors.

Another fundamental difference between an FPGA and a CPLD refers to the storage of the interconnects. While CPLDs are non-volatile (that is, they make use of antifuse, EEPROM, Flash, etc.), most FPGAs use SRAM, and are therefore volatile. This approach saves space and lowers the cost of the chip because FPGAs present a very large number of programmable interconnections, but requires an external ROM. There are, however, non-volatile FPGAs (with antifuse), which might be advantageous when reprogramming is not necessary.

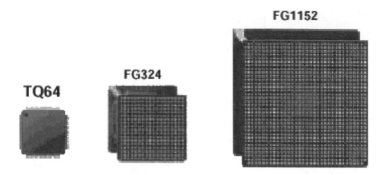

Figure A6
Examples of FPGA packages.

Table A3
Xilinx FPGAs.

Family	Virtex II Pro (X)	Virtex II	Virtex E	Virtex	Spartan 3	Spartan IIE	Spartan II
Logic blocks (CLBs)	352– 11,024	64– 11,648	384– 16,224	384– 6,144	192– 8,320	384– 3,456	96– 1,176
Logic cells	3,168– 125,136	576– 104,882	1,728– 73,008	1,728– 27,648	1,728– 74,880	1,728– 15,552	432– 5,292
System gates		40 k– 8 M	72 k– 4 M	58 k– 1.1 M	50 k– 5 M	23 k– 600 k	15 k– 200 k
I/O pins	204– 1,200	88–1108	176–804	180–512	124–784	182–514	86–284
Flip-flops	2,816– 88,192	512– 93,184	1,392– 64,896	1,392– 24,576	1,536– 66,560	1,536– 13,824	384– 4,704
Max. internal frequency	547 MHz	420 MHz	240 MHz	200 MHz	326 MHz	200 MHz	200 MHz
Supply voltage	1.5 V	1.5 V	1.8 V	2.5 V	1.2 V	1.8 V	2.5 V
Interconnects	SRAM	SRAM	SRAM	SRAM	SRAM	SRAM	SRAM
Technology	0.13 u 9-layer copper CMOS	0.15 u 8-layer metal CMOS	0.18 u 6-layer metal CMOS	0.22 u 5-layer metal CMOS	0.09 u 8-layer metal CMOS		
SRAM bits (Block RAM)	216 k– 8 M	72 k– 3 M	64 k– 832 k	32 k– 128 k	72 k– 1.8 M	32 k– 288 k	16 k– 56 k

Table A4
Actel FPGAs.

Family	Accelerator	ProASIC	MX	SX	eX
Logic modules	2,016–32,256	5,376–56,320	295–2,438	768–6,036	192–768
System gates	125 k–2 M	75 k–1 M	3 k–54 k	12 k–108 k	3 k–12 k
I/O pins	168–684	204–712	57–202	130–360	84–132
Flip-flops	1,344–21,504	5,376–26,880	147–1,822	512–4,024	128–512
Max. internal frequency	500 MHz	250 MHz	250 MHz	350 MHz	350 MHz
Supply voltage	1.5 V	2.5 V, 3.3 V	3.3 V, 5 V	2.5 V, 3.3 V, 5 V	2.5 V, 3.3 V, 5 V
Interconnects	Antifuse	Flash	Antifuse	Antifuse	Antifuse
Technology	0.15 u 7-layer metal CMOS	0.22 u 4-layer metal CMOS	0.45 um 3-layer metal CMOS	0.22 u CMOS	0.22 u CMOS
SRAM bits	29 k–339 k	14 k–198 k	2.56 k	n.a.	n.a.

FPGAs can be very sophisticated. Chips manufactured with state-of-the-art 0.09 μm CMOS technology, with nine copper layers and over 1,000 I/O pins, are currently available. A few examples of FPGA packages are illustrated in figure A6, which shows one of the smallest FPGA packages on the left (64 pins), a medium-size package in the middle (324 pins), and a large package (1,152 pins) on the right.

Several companies manufacture FPGAs, like Xilinx, Actel, Altera, QuickLogic, Atmel, etc. Examples from two companies (Xilinx and Actel) are illustrated in tables A3 and A4. As can be seen, they can contain thousands of flip-flops and several million gates.

Notice that all Xilinx FPGAs use SRAM to store the interconnects, so are re-programmable, but volatile (thus requiring external ROM). On the other hand, Actel FPGAs are non-volatile (they use antifuse), but are non-reprogrammable (except one family, which uses Flash memory). Since each approach has its own advantages and disadvantages, the actual application will dictate which chip architecture is most appropriate.

Appendix B: Xilinx ISE + ModelSim Tutorial

The following synthesis, placement, and simulation tools are described in the tutorials presented in the Appendices:

Tools	Application	Appendix
ISE 6.1 + ModelSim 5.7c	Xilinx CPLDs and FPGAs	B
MaxPlus II 10.2 + Advanced Synthesis Software	Altera CPLDs and some FPGAs	C
Quartus II 3.0	Altera CPLDs and FPGAs	D

Xilinx ISE 6.1 is a comprehensive synthesis and implementation environment for Xilinx programmable devices. ModelSim XE 5.7c (from Model Technology) is also provided as part of the package. The former is employed for circuit synthesis and design implementation, while the latter is used for simulation.

Xilinx ISE 6.1 WebPack, along with ModelSim XE II 5.7c Starter, can be downloaded cost-free from www.xilinx.com.

This is a very brief tutorial, which is divided into five parts:

B1. Entering VHDL Code

B2. Synthesis and Implementation

B3. Creating Testbenches

B4. Simulation (with ModelSim)

B5. Physical Realization

B1. Entering VHDL Code

• Launch ISE 6.1 Project Navigator. A screen like that of figure B1 will be displayed.

• Start a new project (File → New Project). The dialog box of figure B2 will be shown. In the Project Name field, type the name of the ENTITY of the VHDL code to be entered (flipflop, in this example). In the Project Location field, choose the working directory. Finally, select HDL as the top level module type. Click on Next.

• In the dialog box of figure B3, select the device (Spartan 3, for example). Then select XST (Xilinx Synthesis Technology) as the synthesis tool, ModelSim as the simulator, and VHDL as the language. Click on Next.

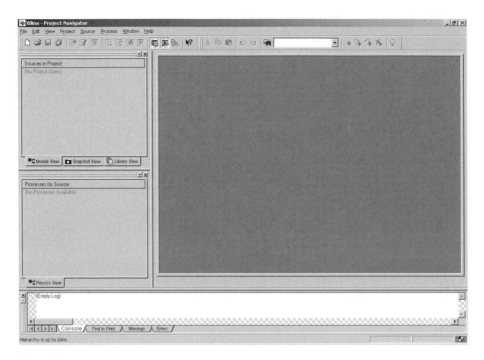

Figure B1

Figure B2

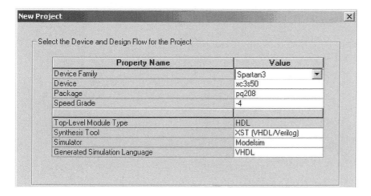

Figure B3

• In the dialog box of figure B4, select VHDL Module, then type the file name (flipflop.vhd, in this example), and choose its location. Click on Next and Finish until the text editor is displayed, as in figure B.5.

• Enter your VHDL code (figure B5) and save it. The project is now ready to be synthesized.

B2.   Synthesis and Implementation

• In the Processes for Source window, select Synthesize-XST. Then go to Process → Properties. The box of figure B6 will be shown. Select Optimization Goal = Area and Optimization Effort = Normal, then click on OK.

• To synthesize the design, select Process → Run, or click on ▣, or double-click on Synthesize-XST. However, if desired, the syntax can be checked before synthesis is invoked. Just click on the "+" sign before the word Synthesize-XST to expand it (see figure B7) and double-click on Check Syntax.

• After synthesis is concluded, view the synthesis report. Double-click on View Synthesis Report, under Synthesize-XST, in the Processes for Source window (figure B7). To better view the report, you can use the toggle tool ▣. A section of such a report is presented in figure B8. Check, for example, the number of flip-flops inferred by the compiler.

• Check also the RTL diagram. Double-click on View RTL Schematic, under the Synthesize-XST directory. The diagram of figure B9 will be presented.

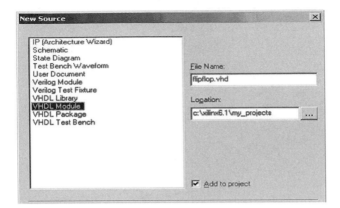

Figure B4

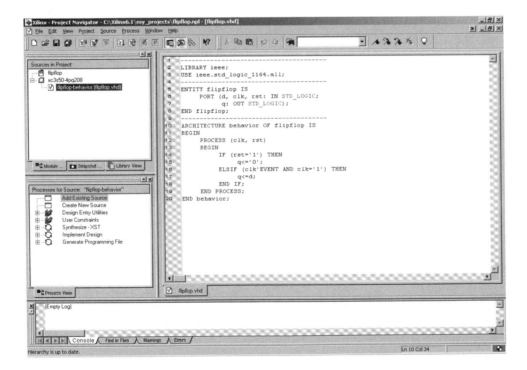

Figure B5

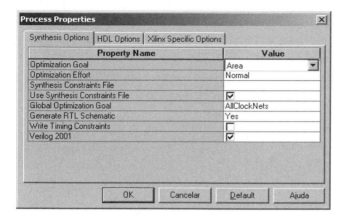

Figure B6

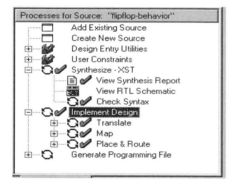

Figure B7

• Now the design can be implemented. Double-click on the Implement Design option in the Processes for Source window (figure B7).

• After the implementation is concluded, expand the Implement Design option and check the several reports produced, particularly the Pad Report (under the Place & Route directory). Check which pin was assigned to each signal.

• Play with the Floorplanner. Double-click on View/Edit Placed Design (Floorplanner), under the Place & Route directory. Select View → Hierarchy, View → Floorplan, View → Placement, View → Package Pins. Now examine each one of windows created. Move the cursor over the pins of the chip to see their descriptions.

Release 6.1i - xst G.23 ================================ Input File Name: flipflop.prj Output File Name : flipflop Output Format: NGC Target Device: xc3s50-4-pq208 Optimization Goal: Area Optimization Effort: 1 Keep Hierarchy: NO Global Optimization: AllClockNets RTL Output: Yes ================================ Synthesizing Unit <flipflop>. Related source file is c:/xilinx6.1/my_projects/flipflop.vhd. Found 1-bit register for signal <q>. Summary: inferred 1 D-type flip-flop(s). Unit <flipflop> synthesized.	HDL Synthesis Report Macro Statistics # Registers: 1 1-bit register: 1 ================================ Cell Usage : # FlipFlops/Latches: 1 # FDC: 1 # Clock Buffers: 1 # BUFGP: 1 # IO Buffers: 3 # IBUF: 2 # OBUF: 1 ================================ Device utilization summary: Selected Device : 3s50pq208-4 Number of Slices: 1 out of 768 0%

Figure B8

Figure B9

Note: Had a CPLD (CoolRunner, for example) been chosen instead of an FPGA (Spartan 3, in this example), the list of options in the Processes for Source window would be a little different. Try, for example, to double-click on the device description (xc3-s50...) in the Sources in Project window. This will bring back the dialog box of figure B3. Change the device to CoolRunner 2. Press OK and then observe the new list of options displayed in the Processes for Source window.

B3.   Creating Testbenches (with HDL Bencher)

HDL Bencher allows the creation of testbenches (waveforms). Then ModelSim can be invoked to perform the actual simulation (ModelSim XE II 5.7c Starter is one of the cost-free third-party softwares provided along with Xilinx ISE 6.1 WebPack).

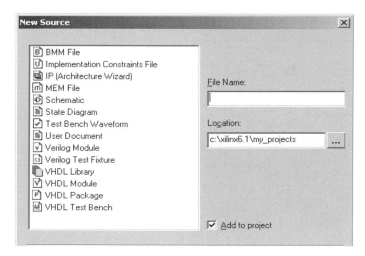

Figure B10

• Select Project → New Source. The dialog box of figure B10 will be displayed. Select Test Bench Waveform, then type the desired file name (flipflop_tbw, for example). Finally, check whether the project location is correct and click on Next until HDL Bencher is launched (figure B11).

• When HDL Bencher starts, a screen like that of figure B11 is displayed, which allows the clock signal to be set. Notice that the input signal clk was chosen as the master clock. Type in its parameters and then click on OK. The waveforms screen shown in figure B12 is then displayed.

• The position of any signal in figure B12 can be changed by just dragging it up or down. Also, if the clk waveform must be changed, click on ⌐₋ or click the right mouse button in the area under the waveforms, which will cause the dialog box of figure B11 to be presented again.

• We must now set up the values of the other signals in figure B12 (rst and d). To do so, just click on the vertical grid line after which you want the value of the signal to be changed. An example, after all input signals have been set up, is shown in figure B13.

• Define the end time of the testbenches. To do so, click the right mouse button in the area under the curves and select Set End of Testbench, then drag the blue line to the desired position.

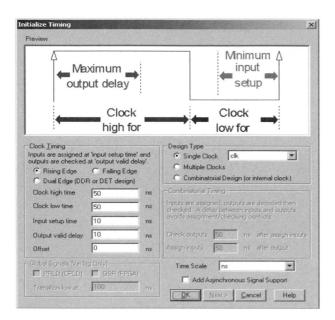

Figure B11

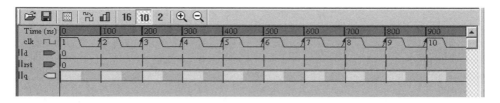

Figure B12

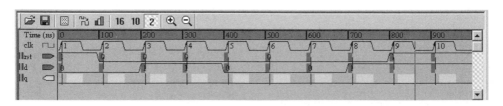

Figure B13

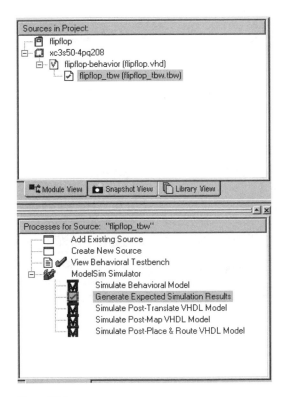

Figure B14

• Save the testbenches file. Observe that a new file (flipflop_tbw.tbw) is then added to the Sources in Project window.

B4.   Simulation (with ModelSim)

Having finished creating the testbenches, ModelSim can now be invoked to perform the simulation. Indeed, several levels of simulation are available, including the following (see the complete list in the lower part of figure B14, under ModelSim Simulator):

• Expected simulation results: Logical verification.

• Behavioral simulation: Logical and timing verification.

• Post-place & route simulation: Logical and timing verification after placement.

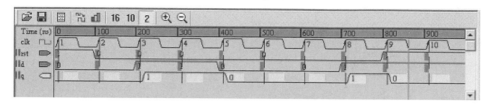

Figure B15

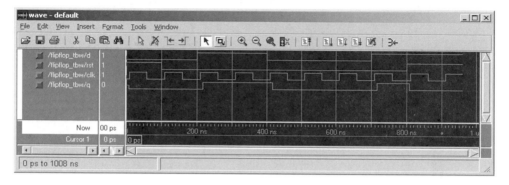

Figure B16

Two of these simulation levels will be employed in the steps below.

• In the Sources in Project window, select the testbench file (flipflop_tbw.tbw). Notice then the several simulation options available under ModelSim Simulator in the Processes for Source window (figure B14).

• Double-click on Generate Expected Simulation Results. This will run a background logical simulator, which will compute the output signals and then automatically launch HDL Bencher with the computed signals included in it. An example is shown in figure B15. Examine whether your project works as expected (from a *logical* point of view). Then exit HDL Bencher without saving the waveforms.

• Now double-click on Simulate Post-Place & Route VHDL Model. ModelSim is launched and a detailed simulation is performed. Maximize the waveforms window and select Zoom → Zoom Full. Examine again the results (figure B16).

B5.   Physical Realization

To physically implement the design in a CPLD or FPGA chip, a development kit is necessary. Inexpensive alternatives are generally available through manufacturer's university programs, which offer design kits at low prices. Xilinx Digilab XC2, for example, is a development kit for Xilinx CoolRunner II devices. The development kit must be connected to a PC running ISE in order for the chip to be programmed.

Since the overall procedure of programming a chip is relatively similar from one manufacturer to another, a detailed description will be presented in only two of the appendices (C and D).

Appendix C: Altera MaxPlus II + Advanced Synthesis Software Tutorial

The following synthesis, placement, and simulation tools are described in the tutorials presented in the Appendices:

Tools	Application	Appendix
ISE 6.1 + ModelSim 5.7c	Xilinx CPLDs and FPGAs	B
MaxPlus II 10.2 + Advanced Synthesis Software	Altera CPLDs and some FPGAs	C
Quartus II 3.0	Altera CPLDs and FPGAs	D

MaxPlus II 10.2 Baseline from Altera is a very simple, user-friendly synthesis and simulation tool. Its main drawback is that it does not support several VHDL constructs, so only relatively simple code can be synthesized without the help of an external synthesis tool (like Leonardo Spectrum or Advanced Synthesis Software). Additionally, it only covers Altera's basic devices (its successor, Quartus II, described in appendix D, covers all current devices). Still, due to its simplicity, it may be an adequate starting point for first-time VHDL users. Moreover, with the recent release of Advanced Synthesis Software, also a cost-free synthesis tool from Altera, using MaxPlus II became more effective because Advanced Synthesis Software does support most VHDL constructs. It can be used to synthesize the VHDL code, generating an EDIF (.edf) file which can then be imported by MaxPlus II for design implementation and simulation.

MaxPlus II 10.2 Baseline and Advanced Synthesis Software can be downloaded cost-free from www.altera.com.

This is a very brief tutorial, which is divided into five parts:

C1. Entering VHDL Code

C2. Compilation

C3. Simulation

C4. Synthesis with Advanced Synthesis Software

C5. Physical Implementation

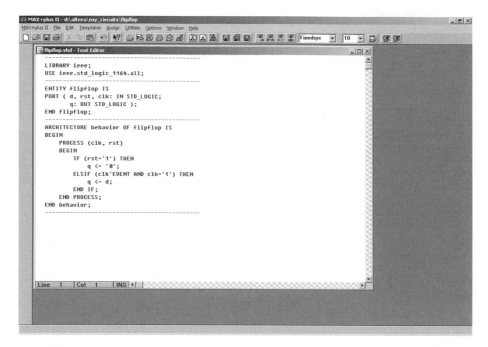

Figure C1

C1.   Entering VHDL Code

· Launch MaxPlus II 10.2 Baseline.

· Open the text editor (MaxPlus II → Text Editor), or open an existing project (File →Open). A blank screen (like that of figure C1, but without the text) will be displayed.

· Enter your VHDL code (a D-type flip-flop is shown in figure C1). Save it with the extension .vhd and using the same name as the ENTITY's (flipflop.vhd, in this example).

C2.   Compilation

· Set the project to the current file: File → Project → Set Project to Current File.

· Choose the target device (Assign → Device). A pull down menu will be displayed (figure C2). Select the desired device (say, Family = MAX3000A, Device = AUTO).

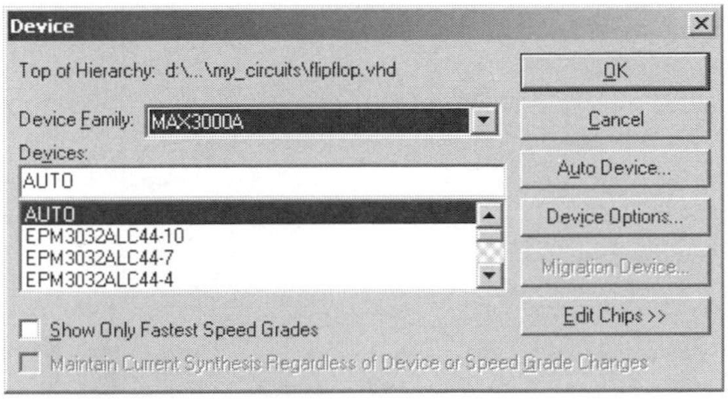

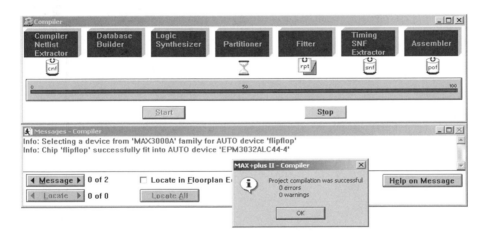

Figure C2

Figure C3

• Set up the optimizer. The implementation can be optimized for speed or for area. Select Assign → Global Project Logic Synthesis and move the Optimize cursor all the way to the left (value = 0) to optimize for area, or all the way to the right (value = 10) to optimize for speed. Values in between can also be used.

• Click on the Compiler icon 🖼, then on Start, in order to execute the compilation.

• If no errors are detected, a screen like that of figure C3 is shown. It displays the files created during the compilation in the upper part (notice, for example, the report "rpt" file icon), and information regarding the chip and fitter in the lower part.

• Open the report (.rpt) file (double-click on its icon, shown in figure C3). Verify at least the following: pin assignments and number of logic cells and flip-flops used to

```
                    R                       R
                    E                       E
                    S         V             S
                    E         C             E
                    R         C             R
                    V    r    I   G  G  G  c  G   V
                    E    s    N   N  N  N  l  N   E
                    D    t  d T   D  D  D  k  D  q D
                 ------------------------------------ _
               /    6   5  4  3  2  1 44 43 42 41 40  |
      #TDI  |   7                                    39 | RESERVED
  RESERVED  |   8                                    38 | #TDO
  RESERVED  |   9                                    37 | RESERVED
       GND  |  10                                    36 | GND
  RESERVED  |  11                                    35 | VCCIO
  RESERVED  |  12            EPM3032ALC44-4           34 | RESERVED
      #TMS  |  13                                    33 | RESERVED
  RESERVED  |  14                                    32 | #TCK
     VCCIO  |  15                                    31 | RESERVED
  RESERVED  |  16                                    30 | GND
       GND  |  17                                    29 | RESERVED
           |_  18 19 20 21 22 23 24 25 26 27 28    _|
                 ------------------------------------
                    R  R  R  R  G  V  R  R  R  R  R
                    E  E  E  E  N  C  E  E  E  E  E
                    S  S  S  S  D  C  S  S  S  S  S
                    E  E  E  E     I  E  E  E  E  E
                    R  R  R  R     N  R  R  R  R  R
                    V  V  V  V     T  V  V  V  V  V
                    E  E  E  E        E  E  E  E  E
                    D  D  D  D        D  D  D  D  D


        Total bidirectional pins required:            0
        Total reserved pins required                  4
        Total logic cells required:                   1
        Total flipflops required:                     1
        Total product terms required:                 2
        Total logic cells lending parallel expanders: 0
        Total shareable expanders in database:        0

              Synthesized logic cells: 0/32   (0%)
```

Figure C4

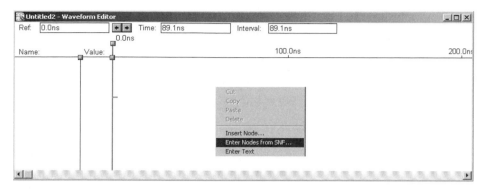

Figure C5

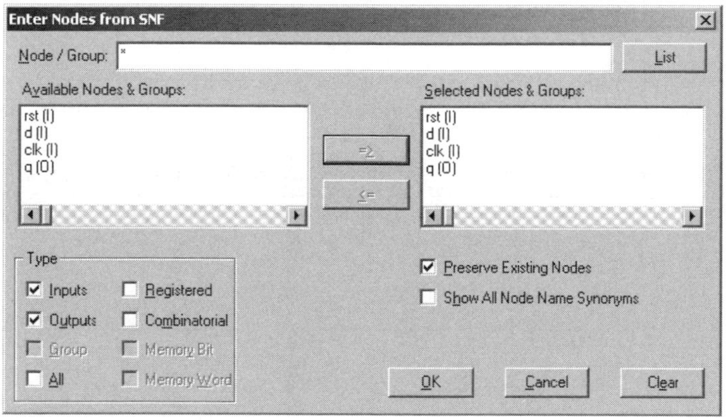

Figure C6

construct the circuit. A little section of the report file from the design of figure C1 is shown in figure C4.

C3. Simulation

• Open the waveform editor (MaxPlus II → Waveform Editor). A blank screen like that of figure C5 will be displayed (without the box in the center).

• With the cursor inside the window of figure C5, press the right mouse button. A pull down menu like that in the center of figure C5 will be shown. Select Enter Nodes from SNF. The dialog box of figure C6 will then be presented. Click on List, then

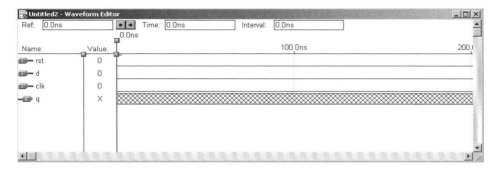

Figure C7

Figure C8

on =>, and finally on OK. All signals listed in the ENTITY of the VHDL code will appear in the waveform window (see figure C7). Notice that the default value for the input signals is 0, while for the outputs it is X (unknown).

• Before establishing the values of the signals, define the length of the waveforms and the grid size. To set the length, select File → End Time and type 1us. To set the grid, select Options → Grid Size and type 50 ns. Finally, select View → Fit in Window. You can also change the order of the signals by just dragging them up or down. For example, to have clk as the first signal, just place the cursor on the arrow that precedes the word clk, then press and hold the left mouse button and drag clk to the desired position. The window will then look like that of figure C8.

• We must now define the input signals, so the tools of figure C9 can be used. The clock icon ▨ is used for pulse generators, ◻ to set the logic value 0, ◻ for logic

Figure C9

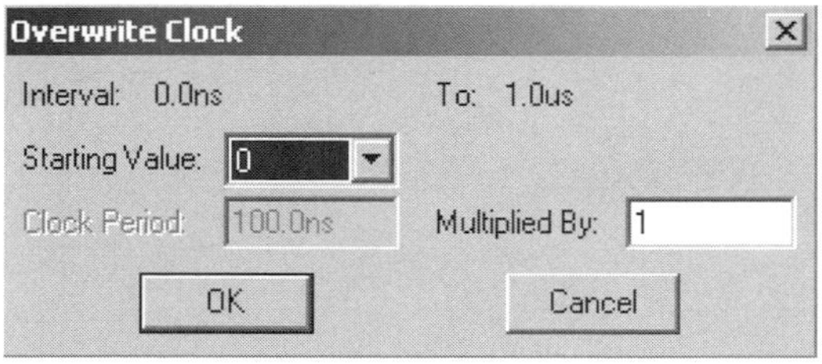

Overwrite Clock   ×

Interval: 0.0ns   To: 1.0us

Starting Value: 0 ▼

Clock Period: 100.0ns   Multiplied By: 1

OK   Cancel

Figure C10

value 1, XC for counters (incremental bus values), and XG for a group value (bus with a fixed value).

• Start with clk. Select the corresponding line (click the left mouse button on the word clk), then click on XO (figure C9), which will cause the dialog box of figure C10 to be displayed. Type Starting Value 0 and Multiplied By 1, then click on OK (Multiplied by 1 means that the period corresponds to one pair of time slots, with each time slot corresponding to one grid space; in this case, period = 100 ns).

• Set up the other input signals. For rst, select the first two time slots (0 to 100 ns). Then click on to change its value to 1 in this interval. Next, select the entire line of d (click the left mouse button on the word d) and click on XO again. Type Multiplied By 4 and click on OK. The waveforms should then look like those in figure C11.

• Save your waveforms with the extension .scf (flipflop.scf).

• Now the design is ready to be simulated. Click on the simulator icon and on Start. The simulator will automatically fill in all output signals in the waveform editor (q, in this example). The result is shown in figure C12.

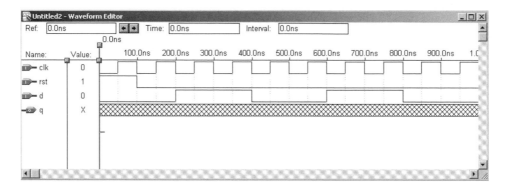

Figure C11

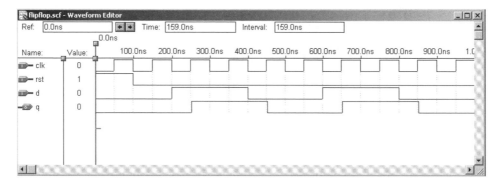

Figure C12

C4. Synthesis with Advanced Synthesis Software

To overcome the limitations of MaxPlus II, which does not support several VHDL constructs, Advanced Synthesis Software was recently released. It can be used to synthesize the VHDL code, giving origin to an EDIF (.edf) file, which can then be imported by MaxPlus II to finish the design (fitting, simulation, programming). As mentioned earlier, Advanced Synthesis Software can also be downloaded cost-free from www.altera.com.

• Using a text editor, type your VHDL code. Suggestion: Since MaxPlus II will be used for fitting and simulation anyway, launch it and type the VHDL code using

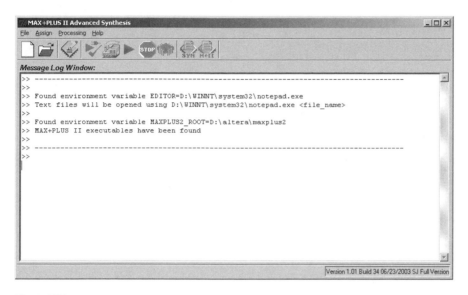

Figure C13

MaxPlus II's own text editor, as described in section C1 above. Save the file with the extension .vhd and the same name as the ENTITY's (flipflop.vhd).

• Launch Advanced Synthesis Software. A screen like that of figure C13 will be displayed.

• Open a new project (File → New Project). In the dialog box, type the name of the project (same as the ENTITY's). The project will be saved with the extension .max2syn (flipflop.max2syn)

• Assign the VHDL file to the project (Assign → Add/remove HDL files). The box of figure C14 will be displayed. Click on Add, select the file, then click on Open and OK.

• Click on the synthesis settings icon ⬚. The dialog box of figure C15 will be presented. Choose the target device (MAX3000A, for example) and VHDL93.

• Click on the synthesis icon ⬚. If no syntax errors are detected, an EDIF file will be generated, with the extension .edf and the same name as the project's (flipflop.edf).

• Return to MaxPlus II and import the EDIF file just created by Advanced Synthesis Software (File → Open). Then start from the beginning of section C2 above, in order to compile the new design.

Figure C14

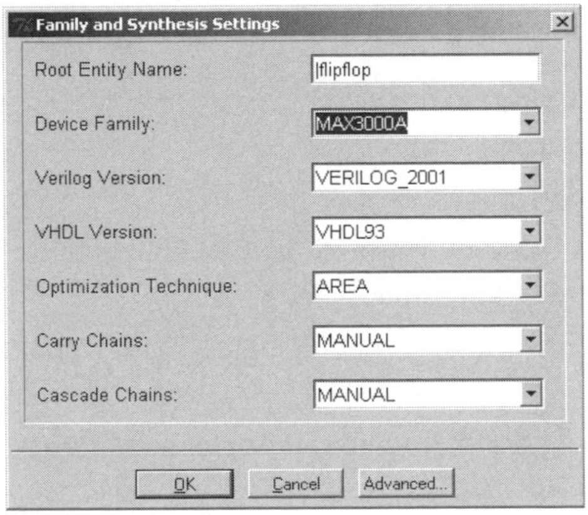

Figure C15

C5.   Physical Implementation

In this section, we will describe the process of physically implementing a circuit on a CPLD. In this description, Altera's UP1 development kit will be utilized, which is furnished as part of their University Program. Other options are also available, either from Altera or other companies. Indeed, most CPLD/FPGA manufacturers offer low-cost development kits as part of their university programs.

The Altera UP1 Board

A view of the Altera UP1 kit is shown in figure C16. As can be seen, it contains two devices:

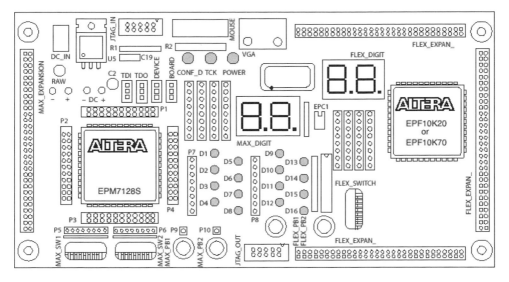

Figure C16

• EPM7128SLC84-7 (from the MAX7000S family): This is a CPLD (appendix A) in an 84-pin package. It contains 128 macrocells, each having a PAL-type architecture and one flip-flop.

• EPF10K20RC240-4 (from the FLEK10K family): This is an FPGA (appendix A) in a 240-pin package. It consists of 1,152 LEs (logic elements), each with a 4-bit LUT (lookup table) and one flip-flop.

For testing the CPLD, the board contains eight LEDs (light emitting diodes), two SSDs (seven-segment displays), and two eight-bit dip switches (figure C16). And, for testing the FPGA, 2 more SSDs and another eight-bit dip switch. The LEDs and the segments of the SSDs use negative logic, thus being turned on when 0 V is applied. The switches, on the other hand, provide 5 V signals when moved up or 0 V when moved down.

The LEDs and switches are not connected to any of the chip pins, so they can be freely wired to the devices to satisfy any particular setup. However, the segments of the SSDs are already connected, thus requiring the implemented circuit to have specific pin assignments. In the case of the CPLD, the pins to which the SSDs are connected are those listed in figure C17.

The board also contains a 25.175 MHz clock, which is connected to the devices (the *global clock* pin of the CPLD is pin 83).

Table 4. MAX_DIGIT Segment I/O Connections		
Display Segment	Pin for Digit 1	Pin for Digit 2
a	58	69
b	60	70
c	61	73
d	63	74
e	64	76
f	65	75
g	67	77
Decimal point	68	79

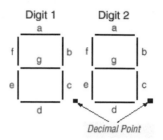

Figure C17

Table 2. JTAG Jumper Settings				
Desired Action	TDI	TDO	DEVICE	BOARD
Program EPM7128S device only	C1 & C2	C1 & C2	C1 & C2	C1 & C2
Configure FLEX 10k device only	C2 & C3	C2 & C3	C1 & C2	C1 & C2
Program/configure both devices	C2 & C3	C1 & C2	C2 & C3	C1 & C2
Connect multiple boards together	C2 & C3	OPEN	C2 & C3	C2 & C3

Figure C18

The complete specifications of the board are available at www.altera.com/ literature/univ/upds.pdf.

Setting up the UP1 Board

• In the description presented below, the CPLD (EPM7128SLC84-7) will be used as the target device. Therefore, the jumpers in the TDO, TDI, DEVICE, and BOARD columns (see figure C16, right above the EPM7128S device) should all be installed in the upper position (that is, between the upper two pins, C1 and C2, of each column of pins, as indicated in the table of figure C18).

• Connect the ByteBlaster cable provided with the kit between the board and the parallel port of the PC.

• Connect the DC supply (9 V) to the board. Notice that the Power LED and two SSDs are lit.

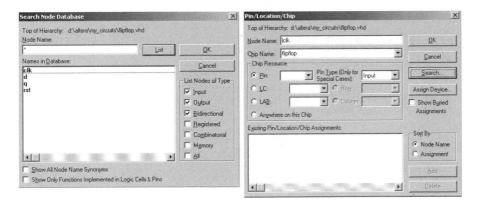

Figure C19

Implementing the Design

We will assume that MaxPlus II 10.2 Baseline is open and that the VHDL code has already been entered and debugged, following the steps described in the previous sections of this appendix.

· Assign the target device by selecting Assign → Device and choosing Family = MAX7000S and Device = EPM7128SLC84-7 (do not check the Select Only Fastest Speed Grade box).

· Compile the circuit as before (click on 🖼).

· Open the report (rpt) file and check which pin was assigned to each signal. If no changes are required, proceed to the next section. To change pins, proceed in the paragraph below.

· To choose a pin for clock different from the automatic global clock assignment (pin 83), first go to Assign → Global Project Logic Synthesis and unmark the box Clock under Automatic Global.

· To choose the pins, select Assign → Pin/Location/Chip → Search → List. A dialog box like that on the left of figure C19 will be displayed. Select a signal and click on OK, thus displaying the box on the right of figure C19. Choose the pin number and the pin type (input, output, etc.), then click on OK if that is the only pin to be changed, or on Add to continue the procedure.

· Upon returning to the main window of MaxPlus II, recompile your design. Then open the report (rpt) file (by clicking on the 'rpt' icon) and confirm that the pins were indeed assigned as expected.

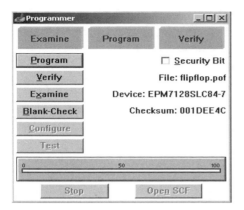

Figure C20

Downloading the Design

Your design is now ready to be downloaded onto the chip.

• Double-click on the pof (program object file) icon (shown at the end of compilation, figure C3). A box like that of figure C20 will be displayed.

• Select, in the main menu, Options → Hardware Setup → ByteBlaster(MV), then click on OK.

• Finally, in the screen of figure C20, click on Program to program the device. After a few moments, the chip will be ready to be physically tested and/or used.

Appendix D: Altera Quartus II Tutorial

The following synthesis, placement, and simulation tools are described in the tutorials presented in the Appendices:

Tools	Application	Appendix
ISE 6.1 + ModelSim 5.7c	Xilinx CPLDs and FPGAs	B
MaxPlus II 10.2 + Advanced Synthesis Software	Altera CPLDs and some FPGAs	C
Quartus II 3.0	Altera CPLDs and FPGAs	D

Quartus II 3.0 from Altera is a comprehensive integrated compiler, placement, and simulation tool. It allows the complete design, from VHDL code to physical implementation, of projects using any of Altera's FPGA or CPLD devices. Quartus II is the successor of MaxPlus II (Appendix C).

Quartus II 3.0 Web Edition can be downloaded cost-free from www.altera.com.

This is a very brief tutorial, which is divided into four parts:

D1. Entering VHDL Code

D2. Compilation

D3. Simulation

D4. Physical Implementation

D1. Entering VHDL Code

· Launch Quartus II 3.0. A window like that of figure D1 will be displayed.

· Create a new project (File → New Project Wizard). The dialog box of figure D2 will appear. Select the working directory in the first field, and the project name (same as the ENTITY's) in the second. The last field will be automatically filled with the project name (you may change it if you want). In the example below, the working directory is d:\altera\my_circuits, and the project name is flipflop. A new project, called flipflop.quartus, is then created in the working directory, which will contain the flipflop.vhd file to be created.

· Open the text editor (File → New, or click on ▣). The menu of figure D3 will then be displayed. Select VHDL File. A blank screen will be presented.

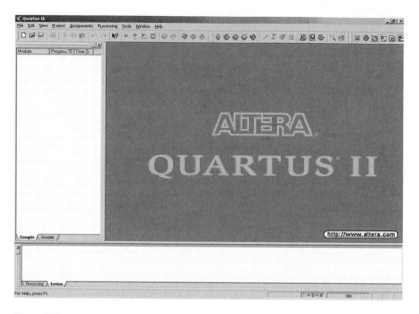

Figure D1

Figure D2

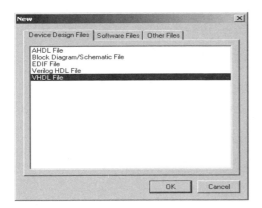

Figure D3

• Enter your VHDL code (as in figure D4). Save it with the extension .vhd (the same name as the ENTITY's will be automatically assigned to the file, that is, flipflop.vhd in this example).

• Check for syntax errors. Select Processing → Analyze Current File, or simply click on the analysis icon ▣. Any error detected by the compiler will be described in the bottom window.

D2.   Compilation

• Select the target device (Assignments → Devices). A menu like that of figure D5 will be displayed. Choose the desired device Family (MAX3000A, for example). In the Target device option, you may select Auto device. In the Package, Pin count, and Speed grade options, select Any.

• To compile your VHDL code, select Processing → Start Compilation, or click on ▶. If successful, a window like that of figure D6 will be displayed.

• Examine the compilation reports (listed on the left of figure D6). Check at least the following:

(a) Flow Summary: This report is displayed automatically at the end of compilation, as shown in figure D6. It contains the part number of the device, the number of pins used, and the usage of the device (number of logic cells used / total number of logic cells).

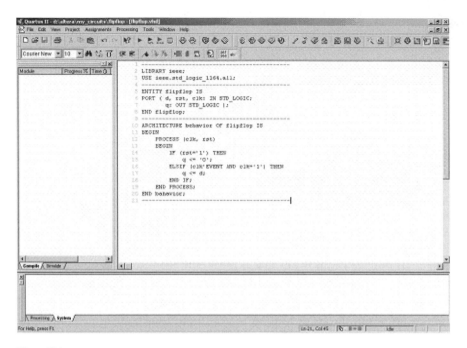

Figure D4

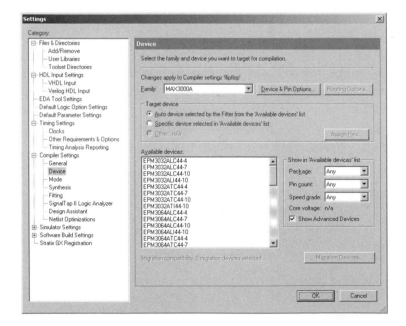

Figure D5

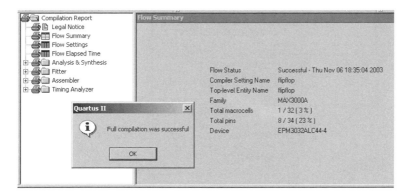

Figure D6

(b) Resource Usage Summary (Fitter → Resource Section → Resource Usage Summary): This report (figure D7) shows details regarding the number of registers inferred from the code, logic cells used, I/O pins, etc.

(c) Input and Output Pins (Fitter → Resource Section → Input Pins, Fitter → Resource Section → Output Pins): These two reports show the I/O pin assignments.

(d) Floorplan View (Fitter → Floorplan View): Shows a layout of the logic cells, which logic cells were used and how, etc. (see figure D8).

(e) Analysis and Synthesis Equations (Analysis and Synthesis → Analysis and Synthesis Equations): Contains the logical equations implemented by the compiler (logical operations + registers).

D3. Simulation

• Open the Waveform Editor. To do so, select File → New → Other File → Vector Waveform File, or simply click on ⬇. A screen like that of figure D9 will be displayed.

• In order to define the size of the waveforms (figure D9), do:
Edit → End Time (select 500 ns, for example).
Edit → Grid Size (select Period = 50 ns, Duty Cycle = 50%).
Finally, select View → Fit in Window.
Note: To change the default values, go to Tools → Options → Waveform Editor → General.

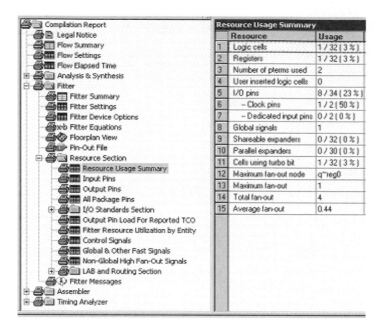

Figure D7

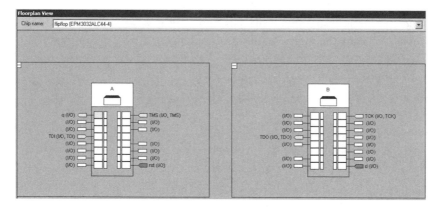

Figure D8

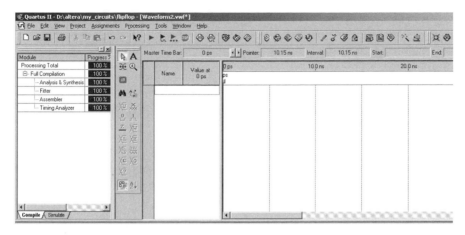

Figure D9

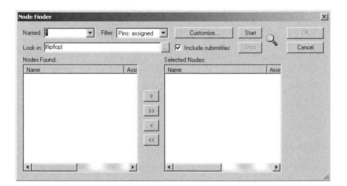

Figure D10

• Add the input and output signals to the waveform window. To do so, click the right mouse button inside the white area under Name (figure D9) and select Insert Node or Bus. In the next box, select Node Finder. A screen like that of figure D10 will then be shown. Make sure that Filter is set to Pins: all. Click on Start, then on ≫, and finally on OK. The waveforms window will now contain a list of all signals described in the ENTITY of the VHDL code, as shown in figure D11. Notice that the input signals (clk, rst, d) are indicated by an inward arrow with an "I" inside, while the output signal (q) is represented by an outward arrow with an "O" inside. The position of the signals can be rearranged by simply dragging them up or down (for example, one might want rst to come right below clk).

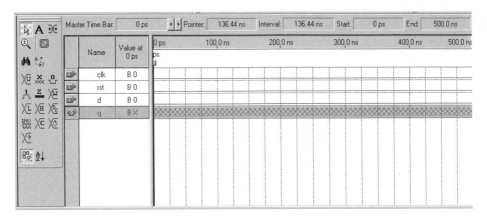

Figure D11

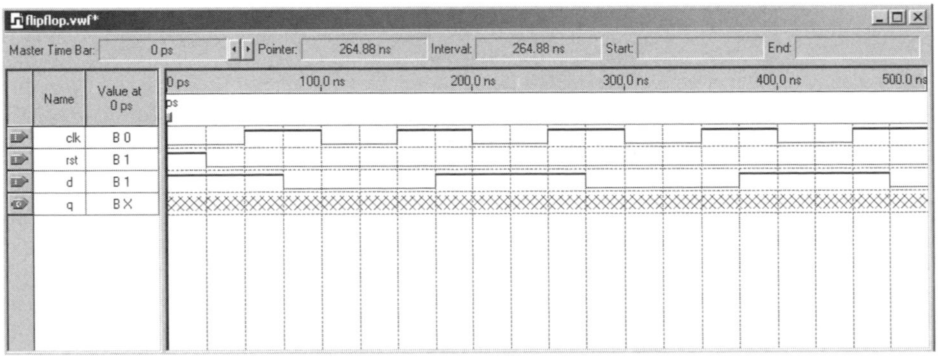

Figure D12

• We have to set now the values of the input signals (clk, rst, and d in figure D11). The easiest way is by using the waveform menu (shown on the left-hand side of figure D11). To set up the clock signal, select the entire clk line (by clicking on the arrow with an I inside beside the word clk) and then click on ✕⊙. A setup box will be displayed. Choose Period = 100 ns.

• For rst, select only its first portion (from 0 to 25 ns), then click on 凡, which will cause the selected portion to change its logic level from 0 to 1.

• Finally, we have to set up the value of d. Select the entire d line, then click on ✕⊙. Choose Period = 200 ns and Phase = 75 ns. The result is shown in figure D12.

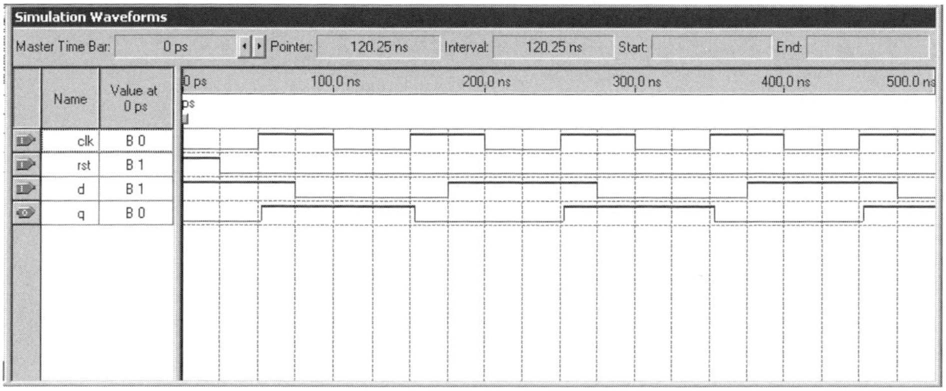

Figure D13

Notice that q is not available yet, for it will be determined by the simulator. Save the waveform as flipflop.vwf.

• The system is now ready for simulation. Select Processing → Start Simulation, or just click on ▶. The result should look like that in figure D13.

D4. Physical Implementation

• Development kit: To perform the physical implementation, we will assume that an Altera UP1 (or UP2) kit is available (this development kit was described in section C5 of appendix C). The kit must be connected to the parallel port of the PC by means of a ByteBlaster cable (provided with the kit).

• Device selection: The kit (Altera UP1 or UP2) contains two devices, EPM7128SLC84-7 (a CPLD from the MAX7000S family) and EPF10K70RC240-4 (an FPGA from the FLEX10K family). Therefore, in the Assignments → Devices step of section D2, one of these two devices must be selected.

• Changing pin assignments: The I/O pins are automatically assigned during compilation. However, if desired, the assignments can be changed. Select Assignments → Assign Pins, which will cause the window of figure D14 to be opened. Say that we want rst to be connected to pin 4, for example. Select pin 4, then click on ▢, which will open the window of figure D10. Click on Start, select rst on the left column, then click on > and OK. Upon returning to the window of figure D14, click on Add. Repeat this process for any other changes of pin assignments.

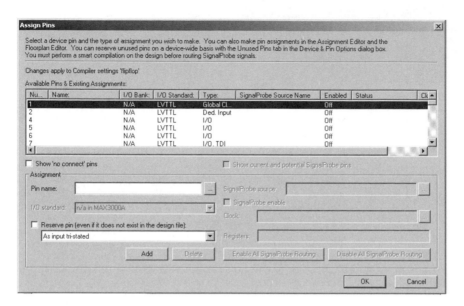

Figure D14

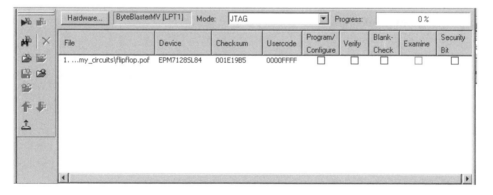

Figure D15

• Setting up the Programmer: To download the program to the kit (device), first select Tools → Programmer, or click on 🖱. The window of figure D15 will be shown. In the Hardware option, ByteBlasterMV (LPT1) should appear. If not, click on Hardware, then on Select Hardware, select ByteBlasterMV, and finally click on Add Hardware. Returning to the window of figure D15, in the File column verify that the design file, with the extension .pof (program object file), is present. Then check the box under Program/Configure.

• Programming the device: Finally, the device can be programmed. Just select Processing → Start Programming. After a few moments, programming will be concluded and the chip ready to be physically tested and/or used.

Appendix E: VHDL Reserved Words

From VHDL 87:	ENTITY	OPEN	WAIT
	EXIT	OR	WHEN
ABS	FILE	OTHERS	WHILE
ACCESS	FOR	OUT	WITH
AFTER	FUNCTION	PACKAGE	XOR
ALIAS	GENERATE	PORT	
ALL	GENERIC	PROCEDURE	*From VHDL 93:*
AND	GUARDED	PROCESS	
ARCHITECTURE	IF	RANGE	GROUP
ARRAY	IN	RECORD	IMPURE
ASSERT	INOUT	REGISTER	INERTIAL
ATTRIBUTE	IS	REM	LITERAL
BEGIN	LABEL	REPORT	POSTPONED
BLOCK	LIBRARY	RETURN	PURE
BODY	LINKAGE	SELECT	REJECT
BUFFER	LOOP	SEVERITY	ROL
BUS	MAP	SIGNAL	ROR
CASE	MOD	SUBTYPE	SHARED
COMPONENT	NAND	THEN	SLA
CONFIGURATION	NEW	TO	SLL
CONSTANT	NEXT	TRANSPORT	SRA
DISCONNECT	NOR	TYPE	SRL
DOWNTO	NOT	UNITS	UNAFFECTED
ELSE	NULL	UNTIL	XNOR
ELSIF	OF	USE	
END	ON	VARIABLE	

Bibliography

Armstrong J. R. and F. G. Gray, *VHDL Design Representation and Synthesis*, Englewood Cliffs, NJ: Prentice Hall, 2nd Edition, 2000.

Bhasker J., *VHDL Primer*, Englewood Cliffs, NJ: Prentice Hall, 3rd Edition, 1999.

Chang K. C., *Digital Systems Design with VHDL and Synthesis—An Integrated Approach*, Los Alamitos, CA: IEEE Computer Society Press, 1999.

Hamblen J. and M. Furman, *Rapid Prototyping of Digital Systems*, Boston: Kluwer Academic Publisher, 2nd Edition, 2001.

Naylor D. and S. Jones, *VHDL: A Logic Synthesis Approach*, London: Chapman & Hall, 1997.

Navabi Z., *VHDL Analysis and Modeling of Digital Systems*, New York: McGraw-Hill, 1993.

Pellerin D. and D. Taylor, *VHDL Made Easy*, Englewood Cliffs, NJ: Prentice Hall, 1997.

Perry D. L., *VHDL*, New York: McGraw-Hill, 2nd Edition, 1994.

Yalamanchili S., *Introductory VHDL from Simulation to Synthesis*, Englewood Cliffs, NJ: Prentice Hall, 2001.

Yalamanchili S., *VHDL Starter's Guide*, Englewood Cliffs, NJ: Prentice Hall, 1998.

Index